After you have read the research report, please give us your frank opinion on the contents. All comments—large or small, complimentary or caustic—will be appreciated. Mail them to CADRE/AR, Building 1400, 401 Chennault Circle, Maxwell AFB AL 36112-6428.

Uninhabited Combat Aerial Vehicles

Airpower by the People, For the People, But Not with the People

Clark

Thank you for your assistance

Cut along dotted line

COLLEGE OF AEROSPACE DOCTRINE, RESEARCH AND EDUCATION

AIR UNIVERSITY

Uninhabited Combat Aerial Vehicles

Airpower by the People, For the People, But Not with the People

RICHARD M. CLARK
Lt Col, USAF

CADRE Paper No. 8

Air University Press
Maxwell Air Force Base, Alabama 36112-6615

August 2000

Disclaimer

This CADRE Paper, and others in the series, is available electronically at the Air University Research web site http://research.maxwell.af.mil under "Research Papers" then "Special Collections."

CADRE Papers

CADRE Papers are occasional publications sponsored by the Airpower Research Institute of Air University's College of Aerospace Doctrine, Research and Education (CADRE). Dedicated to promoting understanding of air and space power theory and application, these studies are published by the Air University Press and broadly distributed to the US Air Force, the Department of Defense and other governmental organizations, leading scholars, selected institutions of higher learning, public policy institutes, and the media.

All military members and civilian employees assigned to Air University are invited to contribute unclassified manuscripts. Manuscripts should deal with air and/or space power history, theory, doctrine or strategy, or with joint or combined service matters bearing on the application of air and/or space power.

Authors should submit three copies of a double-spaced, typed manuscript along with a brief (200-word maximum) abstract of their studies to CADRE/AR, Building 1400, 401 Chennault Circle, Maxwell AFB AL 36112-6428. We also ask for an electronic version of the manuscript on 3.5-inch or 5.25-inch floppy disks in a format compatible with DOS or Windows. Air University Press uses Microsoft Word as its standard word-processing program.

Please send inquiries or comments to:
Dean of Research
Airpower Research Institute
CADRE
401 Chennault Circle
Maxwell AFB AL 36112-6428
Tel: (334) 953-6875
DSN 493-6875
Fax: (334) 953-6739
Internet: james.titus@cadre.maxwell.af.mil

Contents

Photographs

Page

Foreword

In one form or another, unmanned aerial vehicles (UAV) have been employed for over 2,000 years. Lt Col Richard M. Clark's *Uninhabited Combat Aerial Vehicles: Airpower by the People, For the People, But Not with the People,* draws on that long history to gauge what the future may hold for uninhabited combat aerial vehicles (UCAV).

The United States (US) Air Force's experience with UCAVs dates back to World War I and the US Army Air Service's order for 25 Kettering Bugs, explosive-laden unmanned minibiplanes. Over the next 60 years, the Air Force continued to experiment with—and periodically employ—UAVs/UCAVs in peace and war. Operational results were decidedly mixed. The Air Force abandoned UCAV development in the aftermath of the Vietnam War, but by the 1990s there was a marked resurgence of interest in UCAVs as a means of "doing more with less" while reducing combat risks to pilots.

Given the problematic history of UAVs/UCAVs, knowledge of past experience could prove beneficial to the current generation of UCAV developers and planners. To that end, Colonel Clark examines technological obstacles that have handicapped UCAVs historically and which could continue to impede their future evolution. He then turns to more contemporary organizational and cultural issues that might hinder integration of UCAVs into the force. Clark concludes his study by proposing answers to two fundamental questions: (1) What are the major obstacles to UCAVs achieving meaningful operational status in the Air Force, and (2) Can those obstacles be overcome?

Originally written as a master's thesis for Air University's School of Advanced Airpower Studies (SAAS), *Uninhabited Combat Aerial Vehicles* won the 1999 Air Force Armament Museum Foundation Prize as the best SAAS thesis on technology and aerospace power. The College of Aerospace Doctrine, Research and Education is pleased to make this timely study available to the Air Force and beyond.

JAMES R. W. TITUS
Dean of Research
Air University

About the Author

Lt Col Richard Milo Clark, a senior pilot with more than 3,000 flying hours, is a liaison officer assigned to the Legislative Liaison Division, Office of the Secretary of the Air Force, Pentagon. He was commissioned through the United States Air Force Academy (USAFA) in 1986. Following graduation he stayed in Colorado Springs and served as an assistant coach for the USAFA junior varsity football team. After an undefeated season, Colonel Clark attended undergraduate pilot training and graduated in 1988. His first flying assignment was as a copilot on the EC-135 Looking Glass at Offutt Air Force Base (AFB), Nebraska. He was subsequently selected to transition into the B-1B and served as copilot and aircraft commander at McConnell AFB, Kansas, and Formal Training Unit instructor pilot at Dyess AFB, Texas. In 1996 Colonel Clark graduated from the USAF Weapons School. He has a bachelor's degree in management from USAFA, a master's degree in human resource development from Webster University, and a master's degree in national security and strategic studies from the US Naval War College. In 1999 he received a master of airpower art and science degree from the School of Advanced Airpower Studies at Maxwell AFB, Alabama. Colonel Clark was recently selected for a White House fellowship. He and his wife Amy currently reside in Arlington, Virginia.

Introduction

When US Air Force Capt Scott O'Grady's F-16 was shot down over Bosnia in June 1995, Americans watched anxiously as aircraft and helicopters searched for the missing pilot. When O'Grady was retrieved safely from a Balkan forest, television networks cut to special bulletins.

Two months later, an Air Force reconnaissance aircraft also crashed in hostile territory. No attempt to search for the crew was made. The incident rated two lines near the back of most newspapers. Rather than dodging Serbs and eating bugs to survive comfortably, the operators of the Predator unmanned airplane were sitting in an air-conditioned shelter at the USAF's base at Aviano, Italy.

—Bill Sweetman
Popular Science

Bill Sweetman's description of Capt Scott O'Grady's rescue highlights the fact that today's United States (US) military leaders must be sensitive to political and social pressures to keep friendly casualties to a minimum.[1] The loss of a single airman can have a tremendous effect on an entire military operation. Leaders must also contend with shrinking force structures and decreasing military budgets, while the US armed forces remain engaged around the world and across the conflict spectrum. They must find ways to "do more with less." These realities, which are unlikely to change in the foreseeable future, are forcing military leaders to seek new ways to carry on with the business of the United States; and the unmanned aerial vehicle (UAV) is a possible solution to this dilemma.

UAVs are potentially more effective and less costly than manned aircraft because of the removal of the pilot and the associated life support systems. They also remove the pilot from danger, decreasing the risk of casualties and prisoners of war (POW). The US armed forces have used UAVs in past conflicts, but they were used primarily for intelligence gathering. To take full advantage of UAV technology, however, the Air Force is seeking to evolve from the reconnaissance UAV to a multimission uninhabited combat aerial vehicle (UCAV). The pursuit of UCAV technology has proven difficult in the past and will probably be difficult in the future, but the potential payoff makes it a worthwhile quest.

Problem Background and Significance

In the 1960s and 1970s, the Air Force was also engaged in the quest for UCAV technology as a possible solution to the problems of "doing more with less" and reducing the risk to pilots. Toward the end of the Vietnam War, and for a few years after the war, the Air Force looked at using UCAVs to strike enemy targets and for suppression of enemy air defenses (SEAD). The concept never reached fruition, and UCAVs never achieved any meaningful operational capability. The idea was abandoned in the late 1970s.

Recently the UCAV concept was resurrected, and the Air Force is pursuing the capability once again. In order to avoid the same fate that UCAVs met 20 years ago, lessons from the past could prove beneficial. UCAVs have never had a significant operational capability, and UAVs in general have had a tumultuous history. There were several obstacles that hindered UCAVs and UAVs from becoming permanent, significant parts of the Air Force's force structure. Some hindrances were more significant than others, but they all contributed to unmanned aviation's checkered past. Being aware of these obstacles is the first step to overcoming them, and looking at the past can help determine what to expect in the future.

This paper examines the obstacles that inhibited UAVs and UCAVs from achieving significant operational capability throughout history, as well as examines whether these same roadblocks may inhibit UCAVs today and in the future. This paper also seeks to uncover any new obstacles that current UCAV development may face and answers two questions. What are the obstacles to UCAVs achieving meaningful operational status in the US Air Force? Can these obstacles be overcome?

Limitations, Assumptions, and Criteria

The major limitation to answering these questions pertains to determining prevalent attitudes among Air Force decision makers with respect to UCAVs. It is difficult to know exactly how personal biases affect decision making. Because this paper does not actually survey a significant sample of decision makers, this judgement is made based on individual examples of Air Force leaders and their decisions. Even if decision makers were surveyed to determine if personal biases affected their decision to support or not to support UCAVs, it is very possible that many of their answers would not reflect their true attitudes. Most decision makers would like to believe that they are unbiased in their judgements, and many may not realize the extent to which their biases influence their decisions. With these limitations in mind, this paper is as unbiased as possible when assessing the attitudes of Air Force decision makers with regards to UCAVs.

This paper is also limited by the currency of the topic. UCAV technology has recently reemerged, and the Air Force is in the early stages of research and development. Certain aspects such as the costs associated with UCAVs have not yet been released, and this precludes a detailed cost analysis or cost comparison from being included. Costs will be considered only on a very general level, and they will be based on estimates from various sources.

The immaturity of UCAV technology introduces another limitation. A major goal is to anticipate future impediments to UCAV development; however, predicting the future can be problematic. It is difficult to know what technology will be available in 10 or 15 years; therefore, this paper makes the assumption that technology in general will continue to advance as it has in the past. This is an important assumption because the future of UCAVs is heavily dependent upon the technology that becomes available early in the twenty-first century. Specific technologies are discussed in greater detail later, but a certain amount of general technological advancement is required for UCAVs to be used in the future.

In determining what the future will hold for UCAVs, this paper looks to the past. It examines the obstacles that unmanned aviation faced throughout its history to predict what obstacles UCAVs may face in the twenty-first century. Any obstacle that appears to be a hindrance in more than one instance during the evolution of UCAVs will be considered a potential obstacle to future UCAV development. It is not so important to determine which obstacles were most inhibiting to UCAV development in the past, but it is more important to determine the extent to which these obstacles will be factors in the future.

Definitions

A few definitions are necessary. A UAV is defined as a self-propelled aircraft that sustains flight through aerodynamic lift. It is designed to be returned and reused, and it does not have a human on board. This definition excludes lighter-than-air craft such as balloons, blimps, zeppelins, or airships; and it rules out ballistic missiles, which do not employ aerodynamic lift to achieve flight. It excludes cruise missiles. Although cruise missiles are closely related ancestors to UCAVs, they differ because they are one-way platforms, where UCAVs are two-way.[2]

The generic terms *drone, remotely piloted vehicle* (RPV), and *UAV* are used interchangeably; and these types of vehicles fall into three main categories.

- Pilotless target aircraft (PTA) are used to train personnel in air-to-air and surface-to-air target practice. They are also used for the testing of new weapons.
- Reconnaissance UAVs gather intelligence information over enemy territory, and the role of these vehicles is nonlethal.
- Strike UAVs or UCAVs are used as weapons delivery systems to take the offensive against an aggressor with lethal military strikes.[3]

UAVs as target drones and reconnaissance UAVs play a major role in the evolution of UCAVs, and they are discussed. The central thrust is the UCAV.

Though UCAVs are a subset of UAVs, they are differentiated by the fact that a UCAV is a lethal weapon system while a UAV is nonlethal. This is an important distinction. Historically, UAVs have been primarily used for reconnaissance and observation but not for combat operations. Many visionaries see the role of UCAVs extending to operations such as SEAD and deep penetration strikes. In these roles UCAVs would complement manned strike packages. "During the high threat, early phases of a campaign, the UCAV [would] penetrate enemy air defenses and provide preemptive and reactive SEAD and prosecute non-hardened high value targets within the adversary's infrastructure. Throughout the remainder of the campaign, the UCAV [would] provide continuous vigilance and an immediate lethal strike capability to effectively prosecute real-time and time critical targets and to maintain SEAD."[4]

The UCAV would launch from afar and fly to the target area, deliver precision weapons to attack the target, loiter in the area looking for better or additional targets to strike if necessary, and return to base after completing its mission, ready to fight another day.

As mentioned earlier, the ability to return to base after completing the mission to fight another day is the primary difference between UCAVs and cruise missiles. Like manned aircraft, UCAVs will deliver weapons to destroy targets. Cruise missiles, on the other hand, are weapons themselves and do not return to base when the mission is complete. They are good for one-time use only.

Being uninhabited is what differentiates UCAVs from today's manned combat aircraft, but uninhabited does not necessarily mean unmanned. Under most UCAV concepts there is a "man in the loop." This means there is some level of human interface with the system to make decisions at various points in the mission. The man in the loop may operate from a ground station, another aircraft, or a ship; and the amount of interface varies between different concepts. There are some who envision UCAVs as fully autonomous systems where they seek and destroy the target without any human interaction at all. This type of UCAV is truly unmanned and would rely on its own onboard systems, such as automatic target recognition (ATR), to make decisions. The point is that all UCAV concepts call for an uninhabited aircraft, but there are different ideas as to the amount of man-in-the-loop involvement.

Preview of Argument

Though the UCAV concept has received much attention in recent years, the idea is far from new. This paper chronologically traces the evolution of UCAVs beginning two centuries before the birth of Christ (B.C.) and ending with the Air Force's abandonment of UAVs and UCAVs in the late 1970s. This discussion serves two purposes. First, it provides some background information on UCAVs. Second, by looking at the obstacles that prevented past unmanned aviation programs from becoming operationally significant, predictions can be made regarding the obstacles future UCAVs may face.

During the 1980s the Air Force devoted neither time nor resources to UAV and UCAV development, but the 1990s saw a resurgence of activity. This paper examines the most current Air Force involvement with UCAVs. It first looks at the use of UAVs in combat during this decade, and then it describes the current programs established by the Air Force to explore the possibilities of UCAVs.

A discussion follows about the extent to which the obstacles faced by unmanned aviation programs of the past, as described earlier, will hinder UCAVs today and in the future. Past obstacles, however, are only a starting point—new obstacles will also be discussed. Evidence drawn from current periodicals, interviews with UAV and UCAV experts, and other supporting documentation are used to determine what significant obstacles UCAVs may face in achieving operational significance in the Air Force.

This paper concludes by suggesting that these obstacles can be overcome and that UCAV is a technology worth pursuing because the potential payoffs are worth the risks. It also provides recommendations for overcoming the obstacles and managing the risks and uncertainties involved with UCAVs.

Evolution of Uninhabited Combat Aerial Vehicles (UCAV)

The first unmanned flight took place over two thousand years ago when a young man in China stood on a lonely windswept hill and flew "recorded history's first remotely piloted vehicle (RPV)—a kite with a piece of string as a down link to the controller on the ground."[5] The first reference to kites used in a military application was in the second century B.C. when Han Hsin, an ancient Chinese general, used kites to triangulate the distance for a tunnel his army was digging under a besieged city's walls.[6] In 202 B.C. Han dynasty founder Liu Pang surrounded a rival general's forces only to be outwitted by a clever escape tactic. His opponent flew kites armed with windpipes over Liu Pang's forces at night; and the sounds were perceived as supernatural omens of impending doom, causing Liu's forces to flee.[7] In Europe kite flying dates back as far as the second century; but the first military use occurred in 1066 at the Battle of Hastings, where they were used for signaling.[8]

Kite technology advanced rapidly in the nineteenth century. Much of this advancement was due to the work of Sir George Caley, the Father of Aviation, and his work with kite-gliders in 1804.[9] Another significant contributor to early, unmanned aviation was journalist William Eddy. He pioneered the use of unmanned aviation in combat when he took hundreds of Spanish-American War photographs from cameras lifted by kites.[10] Though kites contributed significantly to aviation in general, their main contribution was not their use as unmanned systems but as precursors to manned flight. In 1901 American inventor Samuel Franklin Cody experimented with man-lifting kites and eventually sold a man-lifting reconnaissance kite to the British army.[11]

The first controlled flights, free from the restrictions of kite string, were achieved in the late nineteenth and early twentieth centuries. A pilotless aircraft, built by Samuel P. Langley, achieved the first heavier-than-air, powered, sustained, controlled flight. The steam-powered aircraft was named Aerodrome No. 5 and was launched over the Potomac River on 6 May 1896 for a flight lasting longer than one minute.[12] In Germany another heavier-than-air, powered flight took place in September 1903. Carl Jatho flew an 11-foot, 10-inch-long pilotless biplane powered by a 9.5 horsepower petrol engine over a distance of 196 feet at a height of 11 feet. This flight was earlier and flew farther than the Wright brothers' famous first flight.

It was the Wright brothers, however, who made the most significant contribution to aviation development. They accomplished the first piloted, powered flight on 17 December 1903; and their innovation sparked a technological explosion in aviation that changed the world. But, although unmanned aviation would benefit significantly from the technological breakthroughs made in manned aviation, unmanned aviation would be subordinate to manned aviation from that point on. This fact became evident in World War I.

World War I

On 6 April 1917 America officially entered the First World War. The United States not only sought to enhance its capabilities in manned aviation but it also sought to improve its unmanned aviation force as well. Just eight days after declaring war, the Navy consulting board recommended that $50,000 be allotted to Elmer Sperry's Flying Bomb project. After further investigation of the idea, Secretary of the Navy Josephus Daniels approved $200,000 for the project.[13] Sperry used the N-9 seaplane and a control system that he developed based on his work with gyroscopes to create the Flying Bomb, an early ancestor to modern UCAVs. The maiden flight occurred on 6 March 1918 and was a first for a machine of this kind.[14] The Flying Bomb was envisioned to carry a 1,000-pound bomb load up to 75 miles with an accuracy of about one and one-half miles. The unit cost of the device was predicted to be about $2,500.[15] The program's successes, however, were sporadic; and its lack of progress, coupled with declining funds, led the Navy to cancel it in 1922.[16] Meanwhile, the Army was working to develop its own version of the Flying Bomb.

Nine years after the US Army Signal Corps awarded the Wright brothers a contract for the Army's first manned aircraft, Charles F. Kettering of General Motors was awarded a contract for the Army's first unmanned aircraft.[17] The Army ordered 25 Kettering Bugs on 25 January 1918.[18] The Bug could carry 180 pounds of explosives and cruise at 55 miles per hour with a range of about 40 miles. It was guided to its target by preset controls and had jettisonable biplane wings.[19] Unfortunately, the Bug failed in its testing, only having made eight successful test flights out of 36. The vehicle reportedly needed improvements on the catapult for launching, the engines for power, and the gyros for stability. In the end the project cost the American taxpayer $275,000 and produced no return on the investment.[20]

(Courtesy of the Kettering/GMI Historical Collection)

Kettering Bug

These two World War I projects revealed several problems for early UAVs. First, the experimenters had trouble launching the unmanned aircraft into the air. Second, the manufacturers found it extremely difficult to build a stable aircraft that flew well without a pilot. Limited aerodynamic knowledge, inadequate testing, and hasty construction of the machines caused basic aerodynamic problems with these early, unmanned flying machines. Third, technical problems plagued components such as guidance systems and engines, and this hindered program development. Fourth, the machines were fragile because they were built for one-way missions. Consequently, they were usually destroyed after a crash and this rapidly exhausted the supply for testing. Furthermore, these crashes provided little data for analysis to determine why the crashes occurred. After great expense and effort were put into making these programs fly, very little success was achieved; but this did not prevent unmanned aviation from continuing to develop after World War I.[21]

The Interwar Years

Aviation technology advanced rapidly during the interwar years. For unmanned aviation, the most significant development was radio control. In 1924 the Army Air Corps Engineering Division initiated a program to develop radio controls for unmanned aircraft. In 1928 attempts were made to adapt a commercial Curtiss Robin airplane to carry bombs while being controlled by radio. Radio control also led to the US Army's target drone program. In 1940 the Army Materiel Division began a greatly expanded program to develop a variety of remotely controlled target planes.[22] English actor Reginald Denny, a radio-controlled aircraft enthusiast, brought his hobby to America and opened a model airplane shop on Hollywood Boulevard in the 1930s. In 1941 his company, Radioplane, began providing target drones to the US Army and continued for several years.[23]

Denny's native England was also producing target drones for its armed forces. The DH.82B Queen Bee, designed as a target for antiaircraft gunners, had its first flight in January 1935. It was a biplane made of spruce and plywood and was powered by a 130-horsepower gypsy engine. It could be launched from an airfield or a ship, had a ceiling of 17,000 feet, a maximum range of 300 miles, and a speed of just over 100 miles per hour.[24] The Queen Bee subsequently inspired the US Navy to carve its own niche in radio-controlled aircraft.

In 1936 the US Navy initiated Project Dog, headed by LCDR Delmar S. Fahrney. The project converted single-seat aircraft into UAVs at the Naval Air Factory in Philadelphia, Pennsylvania. The Navy's first radio-controlled, pilotless aircraft, the Curtiss N2C-2, had its first flight in November 1937.[25] Less than a year later, in September 1938, Project Dog unleashed the first assault drone when a pilotless N2C-2 dive-bomber attacked the USS *Utah*. The test was unsuccessful because the aircraft was struck by a stray antiaircraft round from a nearby ship and crashed short of the target, but the program showed enough potential for the Navy to commit itself to the funding and development of assault drones.[26] In April 1941 the Navy conducted the first successful live attack with a remotely piloted Curtiss TG-2. This early UCAV was armed with a dummy torpedo and set out to strike a maneuvering destroyer. Controlled by a "mother" aircraft 20 miles away, the TG-2 released its torpedo and scored a direct hit on the destroyer's target raft. As a result of this and other successful tests, the Navy ordered 500 assault drones and 170 mother aircraft in preparation for the Second World War.[27]

World War II

During World War II, both the US Army and the US Navy dabbled in the use of unmanned strike vehicles. The Army Materiel Division developed several offensive guided weapons in the 1940s. Among these weapons were the GB-1 glide bomb and the VB-1 Azon. The Azon was successfully used against bridges and railroads in Italy; and combat missions were flown out of England with the GB-4, a television-radio controlled glide bomb.[28] The Army Air Forces (AAF) also used remotely controlled aircraft in World War II on a test basis. The program was code-named Aphrodite; and it involved engineers converting "war weary" or fatigued heavy bombers, such as the B-17, into radio-controlled aircraft. They were loaded with more than 18,500 pounds of explosives or napalm and used against high-value, heavily defended German targets. After the pilot got the aircraft off the ground and up to cruising altitude and the technician adjusted the radio equipment and activated the fuze, both men bailed out over England. A control ship using radio control would then guide the bomber to its target; regrettably, none of the Aphrodite bombers were successful. They were either shot down, crashed due to technical difficulties, or simply missed their targets due to the inaccuracy of the navigation system. Gen Carl A. "Tooey" Spaatz canceled the Aphrodite program in January 1945.[29]

At the same time the Army was engaged in Aphrodite, the Navy attempted a similar program that used B-24s instead of B-17s. The Navy's program only consisted of two flights before

(Courtesy of Barnes and Noble Publishing)

Curtiss N2C-2

it was canceled for basically the same reasons as Aphrodite.[30] The Navy did, however, conduct operations with other unmanned assault vehicles that achieved some success. In July 1944 the Navy launched its first operational UCAV missions using TDR-1s, which were drones converted from manned aircraft. Four TDR-1s of the Special Task Group One (STAG-1) launched from the northern Solomon Islands loaded with 2,000-pound bombs scored two direct hits on the Japanese merchantman *Yamazuk Maru*. "STAG -1 launched a total of 46 TDR-1s from Banika Island, near Guadalcanal, between September and October 1944, achieving a 50 percent hit rate."[31] Though there was some limited success in World War II with unmanned aircraft, it was developments after the war that would boost unmanned aviation into a new era.

Post–World War II

Following World War II, emphasis was placed on recycling worn-out, obsolete manned aircraft as aerial targets; but it soon became apparent that an aircraft designed as an unmanned vehicle should be smaller, less costly, and generally more maneuverable than converted manned aircraft.[32] In 1946 the Guided Missile Section of the AAF was formed, and from this section the AAF stood up the first Pilotless Aircraft

Branch. That same year the AAF initiated a project to build two types of target drones. First was the Q-1, a 350 mile-per-hour system. The other was the Q-2, a 600 mile-per-hour machine. Denny's company, Radioplane, won the contract for the Q-1 in 1946; and the Ryan Aeronautical Company was awarded the Q-2 contract in 1948. The Q-2 was to be an aerial target for realistic antiaircraft and air-to-air gunnery practice.[33] This was the first contract for a jet-propelled, subsonic unmanned aircraft, and it would be used by all three services. By the spring of 1951, the first powered flight of the experimental XQ-2 was accomplished, and 32 of the drones were ordered.[34] This was the beginning of a new era. Variants of the Q-2, also known as the Ryan Firebee, would go on to dominate UAV history.

(Courtesy of Barnes and Noble Publishing)

Ryan Q-2 Firebee Target Drone

The Cold War

Early in 1960 Robert R. Schwanhauser, the project engineer for the Ryan Aeronautical Company, briefed Air Force reconnaissance experts at the Pentagon on the possibility of using unmanned aerial drones with a range capability of 1,400 nautical miles for intelligence gathering.[35] Then on 1 May 1960, while gathering intelligence in his U-2 reconnaissance plane,

Francis Gary Powers was shot down over Russia. Due to the political impact of this incident, President Dwight D. Eisenhower believed it necessary to discontinue the U-2 flights. This occurred 18 months before the nation's first reconnaissance satellite would become operational and at a time when work on the higher flying and faster SR-71 had only just begun. "It was also an election year with charges of a 'missile gap' dominating the headlines. At the very time that the 'Cold War' reached new heights, the United States had lost its main source of intelligence behind the Iron and Bamboo Curtains."[36] Not surprisingly, work on an unmanned reconnaissance system that could gather precise photographic intelligence received serious attention. Work intensified on 1 July 1960 when an RB-47 on an electronic intelligence gathering mission over the Barents Sea between Norway and Russia was shot down. Two of the five crew members were taken prisoner in the Soviet Union. Eight days later, a new era of RPV was born.[37]

On 9 July 1960 the Air Force awarded Ryan Aeronautical a highly classified $200,000 contract for a flight-test demonstration showing how its target drones could be adapted for unmanned, remotely guided photographic surveillance missions. The program was code-named Red Wagon, and it was followed by a second program in 1961 code-named Lucy Lee. Unfortunately, both programs ran into higher than anticipated developmental costs and both were terminated.[38] This was a significant setback for the United States's UAV program, but two years later it received new life through the Big Safari procurement concept.

Big Safari was a quick-reaction management concept and would ultimately lead to success for unmanned reconnaissance aircraft. The Big Safari concept was a system of procurement for special reconnaissance that had survived since the early 1950s. "It was an expedited means of bypassing the old research and development system, providing a rapid response capability."[39] Through this program Ryan was granted a contract to modify four standard Q-2C Firebee training targets into reconnaissance UAVs designated as the 147A Firefly.[40]

The Firefly, later renamed the Lightning Bug for security reasons, required considerable alterations to transform it from a target drone into a reconnaissance UAV. The wings were extended from 13 to 27 feet to provide greater altitude capability. The fuselage and nose sections were extended to provide more space for fuel and payload, which allowed increased range and capability to carry intelligence gathering equipment. While the Q-2C training targets were capable of being launched from the ground or air, the 147s were configured only for launch from DC-130s, which were specially modified C-130 cargo planes. After successful test flights, Strategic Air Command decided to update more Q-2Cs for future reconnaissance use.[41] This modification proved to be extremely helpful in preparing the Air Force for the upcoming war in Vietnam.

UAVs in the Vietnam War

On 10 August 1964 President Lyndon B. Johnson issued the Tonkin Gulf Resolution, and the United States became involved in the Vietnam War. This was the first war that would see an extensive use of UAVs. A total of 3,435 operational reconnaissance UAV sorties were flown in Southeast Asia between 1964 and 1975. These sorties involved 1,016 Ryan 147s of varying configurations and models.[42] The 147s made great contributions to the war effort and served as the workhorse of the Vietnam-era UAVs, but Ryan also experienced failure with Vietnam-era UAVs. The Ryan 154 failed to make any meaningful contribution to the war effort and was a setback in the quest to give UAVs a secure position in airpower's operational world.

Success: The Ryan 147 Lightning Bug. The Ryan 147 significantly matured during the Vietnam War, and many improvements and modifications to the system were funded throughout the war. For example, Lightning Bugs were initially designed for recovery by parachute at the end of their mission for reuse. In Vietnam, however, the 147s were badly damaged as they landed in rice paddies, jungles, and the ocean. This prompted Ryan technicians to develop the midair retrieval system (MARS). "After parachute recovery was initiated, a helicopter could snatch the descending bird in mid-air and return it undamaged."[43] This helped to extend the average life of the RPVs used in Vietnam to about three and one-half missions per airframe.[44]

Another program initiated to increase UAV survivability was Operation Chicken. As UAV use increased, so did the number of attempted shootdowns of UAVs. Operation Chicken introduced UAVs to the same tactics used by manned aircraft. This project provided the UAVs with artificial intelligence so they

(Courtesy of the Ryan Aeronautical Library)

Ryan 147 Lightning Bug

could maneuver out of harm's way when threatened. "In all, with the ability to maneuver out of harm's way, the birds out-guessed eight MiG intercepts, three air-to-air missile launches and nine ground-to-air launches. Smart drones were finally out-smarting the smart missiles!"[45] These improvements allowed the UAVs to accomplish a wide variety of missions more effectively, but the primary mission of the 147 in Vietnam was reconnaissance.

Lightning Bugs took photos from high and low altitude with details so clearly decipherable that minute objects could be detected. The photos provided by the 147 were significant throughout the conflict. Photos rendering precise locations of surface-to-air missile (SAM) sites, enemy airfields, ship activity in Haiphong Harbor, and battle damage assessment (BDA) provided valuable intelligence otherwise unattainable unless manned aircraft were used.[46] Manned intelligence gathering missions proved to be very risky for the aircrews flying them.

In June 1969 an Air Force EC-121 Super Connie was shot down off the North Korean coast with 31 men aboard.[47] Its mission was communication and electronic intelligence gathering. This incident emphasized the need for a UAV to do the same job. Six months after the shootdown, the Ryan 147T became operational. Its newly developed engines had 45 percent more thrust than previous models. This allowed for more onboard

(Courtesy of the Ryan Aeronautical Library)

This enemy antiaircraft artillery battery in Vietnam was photographed in 1968 by a Ryan 147 UAV.

electronic gear to listen in on North Korean, Chinese, and Russian communications and to detect radar signals.[48] The TE and TF models were high-altitude UAVs and could intercept signals from target transmitters at ranges up to 600 miles with a relay system to transmit received signals back to ground stations in real time.[49]

UAVs also played a role in the propaganda war against North Vietnam. In July 1972 drones were called upon to deliver leaflets deep in enemy territory. The probability of manned aircraft accomplishing this and returning undamaged was very low, so several 147NC drones were modified with external pods that were usually used for chaff dispensing. In this case, however, they were full of leaflets printed with messages from President Nixon urging the Communists to change their "foolish ways." The success of this project was not earthshaking; and while most referred to the missions as project Litter Bug, the operational troops called the 147NC drones Bulls--t Bombers.[50]

One group that had a special appreciation for the Vietnam-era UAVs was the American POWs in Hanoi. It often provided a great psychological lift for them to see the low-flying vehicles swoop over the "Hanoi Hilton." One returned POW reported, "Sometimes we heard the drone. Sometimes we saw it. After a while the usual comment was 'there goes the little guy.'"[51] Another former POW recalled, "I was standing out in the open in the middle of the compound when a drone approached overhead. I figured it was taking pictures, so I just stood up and smiled for the camera, hopeful that somebody back home might recognize me!"[52] CDR Edward H. Martin, a five and one-half year prisoner stated, "We saw many recce drones during their intelligence gathering flights over the Hanoi-Hilton area. They were the only aircraft we heard for a long time and about the only thing that did lift our morale during those years."[53] Unfortunately, not all of the Vietnam-era UAVs earned the same appreciation as the Lightning Bug did during the war.

Failure: The Ryan 154 Compass Arrow. The photo reconnaissance challenge has always boiled down to flying faster, farther, and more precisely with less vulnerability to enemy weapons while obtaining high-resolution photographs which yield significant intelligence. As the Ryan 147 family of drones opened a new dimension in American reconnaissance, the search was on for an even more advanced system. In 1965 the Air Force determined that there was a requirement for a new, long-range reconnaissance UAV. As the most knowledgeable drone specialist, Ryan—which became Teledyne Ryan Aeronautical (TRA) after Teledyne, Incorporated, purchased it in 1969—was the natural choice to undertake the development of the next generation reconnaissance drone.[54]

The Ryan 154 Compass Arrow went under contract in June 1966 and was a purpose-built reconnaissance drone, as opposed to the 147, which was a modified target drone.[55] The Air Force required the next generation reconnaissance UAV to perform high-altitude, long-range, photographic reconnaissance missions deep in enemy territory. This meant overflying enemy fighters and SAMs. The 154 was designed to fly at 78,000 feet with minimum radar and heat signature. In other words, the 154 was to employ stealth technology.[56] "Using their experience with the Firebees over Vietnam, Ryan engi-

(Courtesy of the Ryan Aeronautical Library)

Two Ryan 154 Compass Arrows are loaded onto a DC-130 "mother" ship.

neers minimized the radar cross-section of the 154 airframe by 'contouring the structural shapes, shadowing the engine intake and exhaust ducts and by using radiation transparent and radiation absorption materials. Infrared suppression [was] achieved by positioning the engine on the top area of the aircraft, extending the fuselage aft of the engine, shadowing the tailpipe by the twin-canted vertical fins and using engine inlet air to cool the engine ejector nozzle.'"[57]

Similar to its Lightning Bug cousin, the 154 was launched from under the wing of a DC-130 "director" aircraft and was recovered by helicopter using the MARS. It could also be recovered by parachute descent to the surface.[58] It carried electronic countermeasures equipment to increase survivability.[59] Because it was a purpose-built reconnaissance UAV with highly classified and sophisticated systems, Compass Arrow was also designed with a built-in self-destruct system.[60]

Approximately 28 of the TRA 154 aircraft were built with the first vehicles coming off the line early in 1969.[61] Unfortunately, Compass Arrow was plagued by several developmental and financial woes throughout its production. "The program quickly fell behind schedule because the project managers had been used to the more loosely administered Big Safari approach [that was used] on the 147 than to the standard procurement methods of the Air Force Systems Command which now had contract responsibility. Several design changes and complexities resulted from building a from-the-ground-up bird, and both Ryan and the Air Force, in time agreed that initially they had been too optimistic in their timetable."[62]

The program was also plagued with control surface actuator problems, MARS recovery system problems, premature engine rundowns, fuel tank leakage, navigational errors, and other chronic malfunctions.[63] TRA and Air Force engineers were able to solve these problems one by one; eventually the 154 was ready to test.

The operational testing of Compass Arrow was conducted between August and December 1971 at Davis–Monthan AFB, Arizona, with launch and recovery accomplished at Edwards Flight Test Center in California. The eight scheduled missions were flown over California and Arizona and lasted 3.4 to 4 hours each. They covered a distance from 1,600 to 1,800 nautical miles at altitudes of up to 81,000 feet at a speed of .8 mach.[64] The final report stated,

> Operationally the [154] is capable of performing high altitude photo reconnaissance missions of 4 hours flight duration, 3.1 hours camera duration, and 1900 nautical miles in length. The Guidance Navigational System (GNS) is capable of executing all programmed functions and directing the Special Purpose Aircraft (SPA) over the planned track with a high degree of accuracy. In addition, because of outstanding vehicle stability, higher photo resolution and improved photo overlap consistency is achieved. Together the [154] and KA-80A [camera system] provide an excellent unmanned strategic reconnaissance system.
>
> GNS performance was excellent and in the primary Doppler/Inertial mode navigational accuracy of less than one-half of one percent of distance traveled can be expected.[65]

Even after the successful operational test and evaluation of the 154, however, the program seemed destined to fail. The capability was present, but it had no mission. The war in Vietnam was winding down, so the 154 would not be needed there. It was designed for use over communist China as well; but owing to the Kissinger–Nixon rapprochement initiative in 1971, it was considered unwise to use Compass Arrow for that mission. The 154 was also suitable for missions in the Middle East during the 1973 Yom Kippur War; but, once again, political considerations prevented its use.[66] For a program that produced little operational capability, Compass Arrow was extremely costly. "The vehicle development program was originally estimated at about $35 million," but the final total system cost was well over $200 million.[67] As one individual who had been involved in the Compass Arrow program put it, "The 154 was an embarrassment to the military-industrial establishment because by the time the money and effort had produced a vehicle of good potential there was no requirement for it, and that was an embarrassment, so the birds were melted down except for the engines."[68]

The Modern UCAV Is Born

Though the 154 Compass Arrow never achieved any operational success, the 147 Lightning Bug made great contributions in the Vietnam War. It was used successfully for a variety of missions, and UAV designers knew of the great potential to expand the scope of missions that UAVs were capable of accomplishing. Though attempts had been made in the past to use unmanned aircraft as strike weapon systems, it was not until 1967 that Ryan first attempted to convert modern UAVs into modern strike weapon systems.

The BQM/SSM Firebee Low-Altitude Ship-to-Ship Homing Missile. Ryan became involved in using UAV technology to deliver weapons in 1967 as a result of the Arab–Israeli Six-Day War, when a Russian-made Styx missile sank the Israeli destroyer *Elath* with the loss of 49 lives.[69] Under the direction of Dr. Robert A. Frosch, assistant secretary of defense for research and development, the development of a ship-to-ship missile, later named the Harpoon, began. The Harpoon would have a range two to three times that of the Styx, but it was still five years before the system would realize an operational capability.[70] Something had to be done in the interim.

For several years, Ryan had been studying and proposing the Firebee low-altitude ship-to-ship homing (FLASH) missile. Unlike solid propellant missiles with a range limited to approximately 35 miles, FLASH would use liquid fuel and have wings, giving it a range of greater than 100 miles.[71] Four capabilities were necessary to convert a standard Firebee (Q-2C) target system into a FLASH weapons delivery vehicle: (1) a demonstrated ability to carry weapons, (2) terminal low-altitude control to the point of contact with the target, (3) the ability to launch the weapon from a ship, and (4) real-time guidance to seek and destroy the target.[72] Ryan was able to demonstrate all four capabilities.

The requirement for the Firebee to carry weapons was not a new idea for Ryan. In the early 1950s, when the Firebee target system first became operational, the Air Force wanted to know what additional uses were practical for the system. Ryan came up with two: reconnaissance and weapons delivery. The reconnaissance mission was inaugurated in 1960, but it was not until late 1964 that Ryan demonstrated the increased payload capability that would make weapons delivery practical.[73] Under US Army Missile Command contract, Ryan conducted a series of ground launches using a large rocket booster. Starting out with two 250-pound bombs, they finally worked up to carrying a one-thousand-pound load of bombs. Using an antisubmarine rocket (ASROC) booster and extended wingtips, the capability to carry weapons was proven.[74]

In response to the second and third requirements, Ryan also relied on previous demonstrations. To prove the Firebee's low-altitude capability, Ryan used the radar altimeter low-altitude control system (RALACS). It had been under development at Ryan as early as 1965 and became operational on a Firebee in 1966. RALACS provided real-time altitude readouts on the remote control operator's console, making it possible to control the aircraft precisely and instantaneously.[75] Ryan demonstrated that the Firebee could be flown under control as low as 50 feet above the surface and at 500 miles per hour. The third requirement, launching a Firebee from a ship at sea, had been previously demonstrated during tests on destroyers and on aviation rescue boats. All test launches had been successful.[76]

The last requirement, that real-time guidance to seek and destroy the target be incorporated into the system, required additional research and flight-testing. The Naval Ordnance Test Station at China Lake, California, was tasked with developing a target seeker and chose to mount a television camera on a Firebee. Several test flights over desert terrain in 1968 successfully demonstrated the concept. The Firebee responded to a proportional control system where the movement of the miniature control stick at the remote control station gave the drone's aeronautical system the same proportional control inputs.[77] "The controller became a ground-based pilot. Watching TV, he could accurately fly the 'Firebee' at low altitudes, just as though he was buzzing the desert in a fighter aircraft."[78] Thus with this successful test, all of the necessary capabilities had been demonstrated.

On 2 September 1971, with several government experts on hand, the BQM/SSM flew a perfect demonstration flight. The Firebee missile was launched from a DC-130 at 5,000 feet; and the remote controller descended the vehicle to 220 feet, then to 75 feet. On final approach it was brought down to 30 feet; and at the last second, the controller dove it to 20 feet. FLASH slammed into the side of the USS *Butler* for a perfect hit. Photo coverage of the impact revealed the ability of the BQM/SSM to demolish any ship of corresponding size.[79]

Although the BQM/SSM proved its worth technologically, it ran into problems with funding because it had to compete with the Harpoon and other missile systems. The Harpoon would be compatible with airborne, surface, and submarine launch platforms; and it would be extremely accurate without having to be monitored by the launch platform after it was fired. The Harpoon also provided an all-weather, antiship capability that the BQM/SSM could not provide.[80] These attributes made Harpoon more attractive than FLASH and led the Navy to aggressively pursue the more capable weapon. During the Harpoon's developmental period, the Navy opted to use existing weapons as interim antiship cruise missiles.[81] This would be less costly than procuring a new system, allowing more funds to be devoted to the Harpoon.[82] The Navy ended the FLASH program without the system ever achieving operational capability, but the program was not a total waste of resources. The BQM/SSM had demonstrated many capabilities that would lead to the use of UAVs as weapons delivery platforms in the future.

The BGM-34A. Like the BQM/SSM, the BGM-34A emerged because of trouble in the Middle East. Israel was concerned about Russian-made SAM and antiaircraft artillery (AAA) batteries, which the Arabs placed along the west bank of the Suez Canal in August 1970. The United States was unwilling to support the Israelis with manned jets to destroy the sites because of the expected high mortality rate.[83] There was additional concern from North Atlantic Treaty Organization (NATO) countries in western Europe because they faced the same threat of multiple SAM sites colocated with AAA sites. "What could President Nixon do to help; what equipment could Washington sell Israel that, if needed, would knock out the SAM and AAA batteries?"[84] On 4 March 1971 TRA was given the go-ahead contract by Air Force Systems Command to develop a UCAV capable of delivering air-to-surface weapons.[85]

TRA engineers drew on the experience and equipment generated over the years to develop the first modern UCAV, the BGM-34A. It was developed in less than one year under the Big Safari management and acquisition program.[86] TRA used the 147S Lightning Bug as the basic frame and combined parts and pieces from six different UAVs to develop the final product.[87] On 14 December 1971 the first full demonstration of the SEAD capability of a drone was accomplished. The BGM-34A was used to fire a powered, guided air-to-surface missile against a simulated SAM site.[88] Gene Juberg, the Ryan engineer leading the test effort, described the flight.

> In the nose was a TV camera equipped with a zoom lens, which in real time transmitted an image of the terrain ahead to the screen in the remote control van. On the drone wing was an AGM-65 Maverick electro-optical seeking missile also capable of telemetering back to control the actual video of the seeker head as it locked on the target. In this case the target was an obsolete radar control van located on the desert to simulate the heart of a SAM site. The remote control operator on the ground was able to see the target on his screen about five miles out without using the zoom capability. At 360 knots airspeed three miles out he switched over to the Maverick's optical seeker and then he was looking at what the weapon was seeing. At about two and a half miles out the missile locked on the target and seconds later it was fired under its own power, hitting the target squarely in the center about nine seconds after firing. It was just as if the ground controller had guided the missile to a direct hit while riding astride the speeding weapon.[89]

It was a historic first missile launching from an RPV to score a direct hit.[90] Seven weeks later an air-launched BGM-34A scored another direct hit when it launched a Stubby Hobo missile, an electro-optical glide bomb with an autopilot to drive the control surfaces for guidance.[91]

At the same time this demonstration of the new defense suppression capability of the UCAV was being tested—December 1971—the bombing of North Vietnam was sharply escalated. The increased number of aircraft used in the bombing resulted in an increase in the number of US planes and pilots that were shot down. "Such large losses added new fuel to the need to develop the unmanned defense suppression capability. In January, the Air Force requested TRA to indicate how many missile carrying UAVs could be rapidly deployed to Southeast Asia."[92] According to Bill Hemlich, a TRA engineer: "The philosophy of Tactical Air Command was to use [UCAVs] to go in on the first wave and soften up the target so that the manned aircraft, F-4 Phantoms and F-105s [could] go in and finish the job with the human eye. We [did not] expect to replace the manned aircraft or the pilot. What we really [wanted] to do [was] go in and soften up the targets and give the pilots a fighting chance."[93]

The Israelis agreed with this philosophy and used the BGM-34A for the delivery of weapons against Egyptian missile sites and armored vehicles. Using the camera mounted in its nose, the BGM-34A located the target and passed it on to the console operator located safely behind friendly lines. After examining the television picture and determining the target's location, the operator selected the target and fired an AGM-65 Maverick missile from the UCAV at the target. The target's image was relayed to another camera housed within the missile, which guided the missile to the target. In December 1973 the Israelis acknowledged their use of UCAVs in the October War.[94]

The BGM-34A was never used in Vietnam, however, because it did not have the technology necessary to perform better than manned aircraft. The SAM sites in North Vietnam were extremely well camouflaged, and the pilots of the manned aircraft that searched for them were unable to find them. It was determined that if the pilots on the scene could not find them with the human eye, then neither could the television camera in the nose of a drone or the electro-optical acquisition and lock-on system in the missile.[95] There was also a need for the development of "a two-way data link that was jam-resistant and covert." This was essential to ensure connectivity between the controller and the UCAV.[96] In a 1972 message from the chief of staff of the Air Force to Tactical Air Command (TAC) headquarters, the Air Force's views regarding the use of UCAVs in Vietnam was explained.

> This headquarters will support all feasible and practical actions for improvement in the tactical drone force and for its establishment as a viable capability. We have not favorably considered the early deployment and use of the TAC drone squadron in SE Asia only because: (A) adequate aircraft and drone assets are not now available; (B) essential modifications to insure satisfactory performance are not complete and tested. . . . We do not preclude use of TAC drone force in SE Asia when an adequate capability has been demonstrated and if it is then required by the combat situation. Request TAC, with assistance as required from [Air Force Systems Command] and [Pacific Air Command], develop a plan for expedited acquisition of necessary aircraft, drone assets and modifications for testing.[97]

(Courtesy of Barnes and Noble Publishing)

A Ryan BGM-34B (background) and a BGM-34A featuring some of the weapons it carried, which included Maverick and Stubby Hobo missiles and Mark 81 and 82 iron bombs.

The Air Force supported the development of UCAVs, but UCAV technology was not sufficiently developed for combat operations. Though UAVs played a significant role in the Vietnam War, UCAVs had no part in that conflict.

BGM-34B. The pullout from Southeast Asia stifled UCAV development. It was two years before TAC was able to resume the development of UCAV capability; but in February 1973, after a one-year-long engineering and product improvement effort, Teledyne Ryan presented the BGM-34B to the Air Force.[98] The BGM-34B differed from the "A" version in that it had a larger more powerful engine, a modified fuselage for increased payload, and enlarged control surfaces to improve aerodynamic capabilities.[99]

The test missions for the new UCAV included single and multiple passes against a plethora of targets. The BGM-34B

(Courtesy of the Ryan Aeronautical Library)

A BGM-34B *(left)* is releasing a self-propelled air-to-surface missile. Configured for strike, a BGM-34B *(right)* is loaded onto a DC-130 mother ship.

was capable of launching a variety of live and inert weapons which included self-propelled air-to-surface missiles and the AGM-65 Maverick television-guided missiles.[100] In November 1974 six demonstration flights were conducted under a US/Federal Republic of Germany (FRG) cooperative project. The objective of the project was to demonstrate to the FRG the feasibility of UCAVs in weather and terrain peculiar to Germany. The missions were conducted within a 30-day period, and all objectives were successfully attained.[101] Unfortunately, the most significant role that the BGM-34B would play was to inspire the development of the BGM-34C.

The BGM-34C. By 1974 Teledyne Ryan had developed UCAVs which could accomplish three major missions: strike, reconnaissance, and electronic warfare. The BGM-34C multimission UCAV, the first of which rolled out for a test in 1976, combined all three capabilities into one platform.[102] Using three interchangeable modular noses and different pods, the basic capabilities of the AQM-34V electronic warfare UAV, the AQM-34M (147SD) reconnaissance UAV, and the BGM-34B air-to-ground strike UCAV were brought together into one UCAV.[103] This concept of multimission UCAVs was expected to lower life cycle and maintenance costs.

Five prototype BGM-34Cs were slated for extensive flight-test evaluations, and future plans called for approximately 20 production vehicles per year starting in July 1977.[104] The 18-

month test program consisted of 27 flights. Each of the different configurations were successfully tested.[105] After the testing was complete, an Air Force news release announced that the next step for the BGM-34C was a production decision by the Department of Defense (DOD). At that time proposals were being considered for the possible procurement of 145 of the new UCAVs over six years. Planned production, however, did not follow. The DOD budget had been significantly reduced after Vietnam, and choices had to be made. The several million dollars slated for the initial production of the BGM-34C systems were diverted to buy tactical aircraft spare parts.[106]

(Courtesy of the Ryan Aeronautical Library)

Capable of strike, reconnaissance, and electronic countermeasures missions, the BGM-34C is shown with a variety of nose and pod configurations.

The End of an Era

After the official cease-fire in Vietnam, reconnaissance drone operations were put on hold. There were a few follow-up validation reconnaissance and communication intelligence flights conducted through the summer of 1975, but there was a serious reduction from what had gone on in the previous years of the war. There were also some updating programs, such as BGM-34A, B, and C strike UCAVs, which were undertaken to refine state-of-the-art technology. In 1976 all UAV programs came under control of TAC. By then the crisis in

Southeast Asia had cooled down, and this led to new constraints on the defense budget.[107]

The Air Force had to make choices regarding UAVs in general because they competed with systems such as the high-speed antiradiation missile (HARM), the B-52/cruise missile force, and tactical strike aircraft for the limited dollars available.[108] Despite presentations made to TAC by TRA president Teck A. Wilson, no new funding was provided to maintain the UAV capabilities which had been developed over the previous 12 years.[109] With the sharp defense budget cuts in the mid-1970s, priorities had to be established. In 1979 more than 60 air-launched recoverable UAVs and UCAVs in various configurations were sent to the mothball fleet.[110] The end of an era had come, and it would be a decade before the Air Force engaged in any more UAV or UCAV activity.

Obstacles along the Way

The evolution of UCAVs was like the movements of the tide, constantly ebbing and flowing. Between World War I and the end of the Air Force's involvement with UAVs and UCAVs in 1979, there were numerous hindrances that caused the ebbs in the UCAV evolution.

Technological Deficiencies. There were several instances where the technology was not adequate enough to support the concept. The Sperry Flying Bomb and the Kettering Bug are two examples of World War I programs that were canceled because the technology had not yet caught up with the concept. Both programs were unsuccessful in their testing, neither was used in the war, and both were canceled before deficiencies could be corrected. The Aphrodite program in World War II had similar problems. The lack of technical know-how led to poor accuracy, extreme vulnerability to enemy defenses, and numerous crashes due to technical difficulties. Finally, the BGM-34A was not used in Vietnam because it was unable to locate camouflaged targets with its built-in television camera or the electro-optical seeker on its missile. This was at best only as good as the human eye, and it was unlikely that these UCAVs would perform better than manned aircraft. There were also data-link and command and control (C^2) technologies that were not mature enough for UCAVs to become operational in the war. In each of these cases, UCAVs were prevented from making significant operational contributions in wartime because of technological challenges.

Managerial Impediments. The most successful UAV in history was the Ryan 147. It was procured under the classified special management program Big Safari. This classified, shortcut, acquisition approach kept the tasks and capabilities of its programs highly classified, making management simpler and less cumbersome. The BGM-34A was also managed under this system, and it was developed in less than a year. Unfortunately, the Big Safari approach was not used for subsequent UAV and UCAV projects; and their development suffered. It was more drawn out and restrictive to manage a program under the traditional Air Force procurement system. A prime example of this difficulty was the Ryan 154. It was managed under the traditional procurement system, and it was plagued with managerial problems that included being behind schedule and over budget. These problems contributed to its cancellation. Subsequent programs also suffered because the secrecy of the Big Safari program left behind insufficient documentation to support future UAV and UCAV programs. In 1972 "the Air Force Inspector General noted the difficulty in establishing a normal management function for the follow-on UAVs since no procurement and engineering data was purchased under 'Big Safari.'"[111] These types of management problems not only hampered the development of UAVs and UCAVs but they made them appear less attractive to civilian leaders.

Political Reluctance. Many congressional members voiced support for UAVs in the mid-1970s. In his remarks to the House of Representatives in 1974, Barry J. Shillito, former assistant secretary of defense for installation and logistics, stated, "Practically every member of Congress is completely sold on the future and, most importantly, the economic need for these type vehicles."[112] In 1974 the Senate Armed Services Committee received two complete briefing on RPVs. One supporter, Sen. Thomas McIntyre of New Hampshire, asked whether the services were devoting enough resources to "this very interesting field." He also stated, "This is an area that we should be pursuing vigorously. . . . Keep looking at it long and hard."[113]

Despite these encouraging words, however, political leaders did not give UAVs commensurate support. UAV programs received significant budget cuts in the 1970s. In Mr. Shillito's opinion, UAVs received little funding because "The public was unaware of their role and only a very few persons in government were permitted to know of their activities. Their comparatively insignificant funding, plus this very recent awareness of their capabilities, undoubtedly is the reason Congress has not been able to give much attention to this very small segment of the DoD budget."[114] In reporting on the fiscal year 1974 DOD appropriations bill, the House Committee on Appropriations reduced the Air Force's $8,400,000 UAV request to $5,000,000. The committee gave the following explanation for the cuts:

> All of the military services are independently pursuing the development of remotely piloted vehicles and drones. There have been some efforts to coordinate these programs and to produce drones which will be utilized by all three services, but for the most part the effort has been unsuccessful. . . . The hearings [also] pointed out that the Air Force spent $250 million to develop a high-altitude reconnaissance drone known as "Compass Arrow" [the Ryan 154], and that after the vehicle was developed, it was placed in dead storage and never flown operationally. In view of past waste in this area and in view of the resistance of the services to full cooperation in a tri-service effort in drones and remotely piloted vehicles, it is recommended that funds be curtailed until realistic overall Department of Defense requirements and systems can be formulated.[115]

A lack of UAV exposure to congressional members, minimal return on dollars invested in UAV research and development, and lack of service cooperation in the pursuit of UAVs were all major causes for the political reluctance to fully support UAVs.

Lack of Service Cooperation. The appropriations committee's remarks about the services' lack of cooperation on unmanned aviation technological development seems to be a recurring trend throughout the evolution of UCAVs and an impediment to success. There were instances in which the services worked parallel programs, and instead of attempting to pool their resources for a single successful program, they both failed. During World War I, the Navy worked to develop Sperry's Flying Bomb while the Army invested resources in the Kettering Bug. The programs were very similar, and both failed due to lack of funds and poor performance. A joint effort on either of these projects could have led to a different outcome and could have resulted in a lower expenditure of resources. In World War II, the Army and Navy duplicated a failing effort with the Army's project Aphrodite and the Navy's similar program using B-24s instead of B-17s. The programs were almost identical, except for the airframes, but there was no evidence of cooperation. A pooling of resources may have made a difference. One program that demonstrated what cooperation among the services can achieve is the Ryan Q-2 target drone. The Q-2 program was a tri-service effort, and it produced the base airframe for the most successful and extensively used UAVs in history. The cooperation among the services on UAV and UCAV development was not only attractive to political leaders but could have helped some early UCAV projects achieve operational success.

Pro-Pilot Bias. It is probably safe to say that when the first man piloted a kite over the hills of China, a better unmanned flying machine was not his ultimate vision. Like most men enthralled by aviation, he probably envisioned manned flight as the supreme goal; and little did he know that his efforts were paving the way for that dream. Two thousand years later, men are still enamored with flight; and in the 1970s, there were many who believed that this attitude was the primary obstacle to RPVs becoming operationally successful. In one journalist's opinion, "within the Air Force, the RPV presents something of a cultural problem. Pilots view the vehicles as a direct challenge." He quoted one Air Force official who proclaimed, "How can you be a 'tiger' sitting behind a console."[116] He also reported that a pilot's "professional sensitivity" was a factor in retarding Air Force interest in UCAVs.[117] When asked about the main problem in selling the UAV idea to the Air Force, Schwanhauser, vice president of Teledyne Ryan Aeronautical, replied that "all the customers wear wings." He implied that pilots were not interested in pilotless airplanes.[118] One researcher studying UCAVs in 1975 deduced that "as an indication of how sensitive this issue is, the report of a study designed specifically to define man's role in the control of strike RPVs classified the discussion of whether or not the operator should be a pilot."[119]

There is, however, an alternative reason for the reluctance by the Air Force and its pilots to invest in unmanned aviation. In 1975 Undersecretary of the Air Force James W. Plummer cautioned government and industry representatives attending the symposium of the National Association of Remotely Piloted Vehicles to temper their enthusiasm with respect to UCAVs: "I think we learned some lessons [in the past.] We are going to have to be convinced of the operational utility of a system before we initiate a full-scale development program, even if the program cost is projected to be small. Further, we are going to be cautious about initiating a vehicle development program where we don't have a good handle on the technological status and requirements of the support system. We cannot justify spending money to prove a concept which may have marginal utility."[120]

These statements indicate that perhaps the reluctance of the services to embrace UCAVs was not based on the threat to the status of pilots and manned aircraft but on the Air Force leadership's skepticism towards the effectiveness of UCAVs. This skepticism may indicate that within the Air Force culture there is an aversion to taking risks on unmanned aviation technology until the uncertainty is reduced significantly. Whether the reluctance towards unmanned systems is based on risk aversion or based on the maintenance of the status of pilots, it is an obstacle inherent in the Air Force's culture that must be addressed.

Competing Weapons Systems. Another obstacle that UAVs and UCAVs had to face in the past was competition for funding from other weapons systems. Manned aircraft were an obvious competitor. For example, the BGM-34C UCAV was cut after completing a successful test program to free up money for manned tactical aircraft parts. This action showed that UCAVs were far below manned aircraft on the budgetary priority list and would have a difficult time competing for dollars. Another competitor was cruise missiles. Two examples show where UCAVs stood against cruise missiles. The BQM/SSM antiship weapon was cut because the US Navy preferred the Harpoon cruise missile. Though it was not a returnable, reusable UCAV, it had proved its capability and could have been an effective weapon; but the Harpoon, which was still under development, won out. The second example is the major cuts that all UAV programs suffered in 1976. These cuts were made because the Air Force leadership invested in other weapon systems such as the B-52/cruise missile force. In those times of tight budgets and reductions in military spending, priorities had to be made. UCAVs took a backseat to the competition: manned aircraft and cruise missiles.

Poor Cost-Effectiveness. Throughout their evolution, several UCAV and UAV programs have been plagued by high costs with minimal returns. Aircraft such as the Sperry Flying Bomb, the Kettering Bug, and the Ryan 154 Compass Arrow and programs like Red Wagon and Lucy Lee all had large amounts of money invested in them, but produced little in return. If these systems had come on-line, would they have been worth the investment? Undersecretary Plummer answered no in 1975. Plummer said, "The initial enthusiasm engendered by this 'lower cost' idea, perhaps, has been over-publicized and the 'dollar saving' potential improperly interpreted. In general, I do not believe that we have met the challenge of adapting technology to produce cost effective RPV systems." Plummer continued, "The hard core issue is whether RPVs can perform traditional missions and save dollars." He also confessed, "We have become over-enamored with our RPV successes." As an example, he brought up the fact that in 1965 the United States set out to develop the high-altitude Ryan 154 UAV from scratch. It was to cost $35 million but ended up costing more than $200 million and was obsolete before an operational mission was ever flown. Plummer believed this was an important lesson.[121]

Is There a Need? The development of UAVs and UCAVs thrived when the need for them was apparent; but when there was no need, they were forsaken. When the United States entered World War I, the Army and Navy were both anxious to develop their versions of the Flying Bomb. When the war was over and the Flying Bombs did not test successfully, interest faded quickly. Though the Navy dabbled in UCAVs during the interwar years, it did not seriously invest in UCAVs until the Second World War broke out in Europe. The shootdown of Francis Gary Powers provided a tremendous boost for modern UAVs, but it was the Vietnam War that generated the greatest requirement. Once the Vietnam War ended, however, the Air Force's interest in UAVs and UCAVs began to dwindle. The 154 Compass Arrow and the BGM-34 series of UCAVs were terminated as the Vietnam War wound down. By 1979 all Air Force UAV and UCAV programs were terminated. For the next decade UAVs and UCAVs lay dormant in mothballs. Unlike manned aircraft, which received constant dollars in war and peace, unmanned aircraft received little or no attention when there was no immediate need for them.

Now that several obstacles to UCAV development have been gleaned from the past, these obstacles can be used to look at the future of UCAVs in the United States Air Force (USAF). Before looking into the future, however, the next section briefly examines more recent UCAV activity. Though UAV activity was nonexistent in the Air Force during the 1980s, it received significant attention during the 1990s. This section reviews the operational activity of UAVs in US armed conflicts in the

1990s, and it discusses the efforts of the Air Force in terms of UCAV development in the last decade of the twentieth century.

Today's UCAVs

Two major operations in which the US armed forces were involved in the 1990s were the Persian Gulf War in Southwest Asia and the peacekeeping operations in Bosnia. Though UCAVs were not used in either of these operations, UAVs were. The use of UAVs in these conflicts was significant in the advancement of unmanned aviation.

Persian Gulf War

Though there was fairly extensive UAV use in the Vietnam War, general awareness of the value of UAVs for military operations did not emerge until Operations Desert Shield and Desert Storm. The US Army, US Navy, and US Marines successfully used UAVs in the Gulf War to contribute to their tactical successes. During the conflict the majority of US manned tactical reconnaissance assets were committed to action throughout the theater, allowing UAVs to emerge as a "must have" capability. For example, the Navy used Pioneer UAVs launched from battleships to support shore bombardment operations, and the Army used the Pioneer UAV for target designation, damage assessment, and reconnaissance.[122] The Army also used the Pointer micro-UAVs, but poor weather and high winds made it less effective than the Pioneer.[123]

The Marines faced a particularly serious tactical intelligence shortfall because their RF-4B reconnaissance aircraft had been

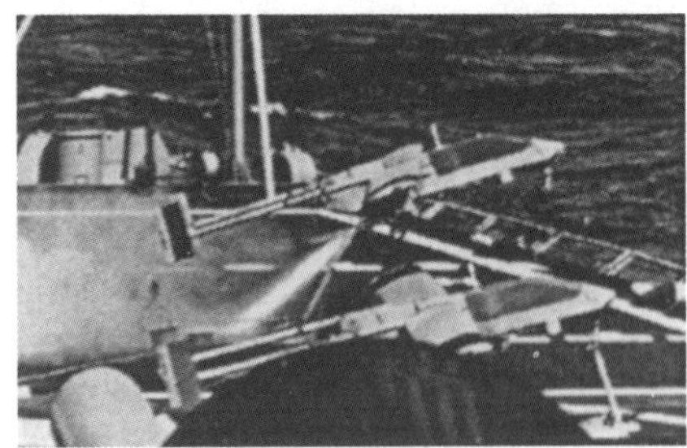

(Courtesy of Barnes and Noble Publishing)

US Navy Pioneers launch from a warship using rocket assisted take-off *(left)*, and a US Marine Pioneer is ready for launch from a mobile catapult ramp *(right)*.

taken out of service before the war and had not yet been replaced.[124] One account described how "UAVs were used to map Iraqi minefields and bunkers, thus allowing the Marines to slip through and around these defenses in darkness, capture key command sites without warning, and speed the advance into Kuwait City by as much as two days."[125] The attack on the Iraqi-held Kuwaiti airport provides another illustration of the utility of UAVs. During that encounter "a live Pioneer UAV picture showed a battalion of Iraqi tanks poised on the north end of the airfield for a counterattack. The armored force was broken up by naval gunfire and air attacks before it could strike the advancing Marines."[126] In one instance Iraqi soldiers surrendered to a Marine Pioneer during battle in Kuwait.[127]

Overall, DOD's final report on Desert Storm concluded that UAVs "proved excellent at providing an immediately responsive intelligence collection capability."[128] They provided highly valued near-real-time reconnaissance, surveillance, target acquisition, and BDA throughout the conflict.[129] "The Gulf War proved to be an important watershed in UAV operations and directly led to the development of Predator, DarkStar and Global Hawk."[130]

(Courtesy of Barnes and Noble Publishing)

Predator Medium-Altitude Endurance UAV

Bosnia

With the success of UAVs in the Gulf War, they have become a popular platform for aerial surveillance over Bosnia and elsewhere in the Balkans. In an effort to decipher the complex situation in the region, the United Nations and NATO deployed a plethora of surveillance assets, including UAVs. The first UAV deployed over Bosnia was the Gnat-750 long-endurance UAV. The Central Intelligence Agency operated it from the Croatian island of Hvar in 1993. Shortly after the Gnat-750 became operational in the region, the US Marine Corps deployed Pioneer UAVs with Task Force Eagle, the US contingent to the intervention force (IFOR). In July 1995 the Predator UAV deployed to Albania for a six-month joint service operation during which Predators flew 128 missions in support of NATO operations Deny Flight and Deliberate Force. USAF Predators operated from Tazsar, Hungary, for another six-month deployment in support of IFOR. At least two Pioneers and three Predators have been lost over Bosnia since operations began. Most of the losses were due to mechanical failure.[131]

The Air Force and UCAVs in the 1990s

The success of the UAVs in the Gulf War and Bosnia generated significant attention for unmanned aviation. It led many planners and developers to consider roles other than reconnaissance and surveillance that UAVs could accomplish in combat, and one of those roles is the SEAD/strike mission for UCAVs. The Air Force has taken active measures in determining whether the UCAV is a concept worthy of investment; and this section examines two of those measures—the UAV Battle Lab and the UCAV ATD program.

UAV Battle Lab. On 1 April 1997 the Air Force opened the UAV Battle Lab at Eglin AFB, Florida. Battle lab members are pilots, intelligence officers, and other specialists charged with exploring the future of UAVs and UCAVs. The battle lab demonstrates UAV and UCAV capabilities, reports on findings, and makes recommendations to the corporate Air Force as to what actions should be taken regarding the capability.[132] Even though the use of UCAVs in combat is many years away, the battle lab is the first step towards establishing requirements. The battle lab is, however, concerned about more than just the technology of UCAVs. The members concentrate on other aspects such as C^2 considerations, integrating the UCAV with other assets in the battlefield, and formulating international agreements for the use of UCAVs. It is their job to look at the practical application aspects of UAVs and UCAVs.

UCAV Advanced Technology Demonstration Program. In 1997 the DOD took another step and created the advanced technology demonstration (ATD) program under the auspices of the Defense Advanced Research Projects Agency (DARPA) and the Air Force. It remains today the principal DOD program exploring the technological aspects of UCAVs. The goal of this program is for the Air Force and DARPA to work together to "demonstrate the technical feasibility for a UCAV system to effectively and affordably prosecute 21st century SEAD/strike missions within the emerging command and control architecture."[133] The UCAV ATD program's primary objective is "to design, develop, integrate, and demonstrate the critical technologies pertaining to an operational UCAV system. The critical technology areas are command, control, and communications; human-systems interaction; targeting/weapons delivery; and air vehicle design."[134] Another key objective for the UCAV ATD program is to "validate a UCAV weapon system's potential to affordably perform SEAD/strike missions in the post-2010 time frame. Life-cycle cost models will be developed which include verifiable estimates of acquisition and operation and support (O&S) costs. The critical affordability assumptions and technologies will be validated through concept and process demonstrations."[135]

The ATD program is divided into two phases. In phase one industry teams participated over a 10-month period to develop proposals in three major areas: a UCAV operational system (UOS), risk-reduction activities, and a UCAV demonstrator system (UDS). DARPA and the Air Force provided the contractors with a proposed mission set for the UCAVs to accomplish. The mission set was defined using state-of-the-art simulation tools; and its design was threat driven, where the missions were based on a simulated threat in the 2010 time frame. The industry teams used the mission sets to design a UOS that could affordably and effectively defeat the future threat. Based on that UOS design, the industry teams determined the critical technologies, processes, and system attributes that a UCAV would have to possess to be effective. Each team listed those technologies, processes, and attributes that were immature and then developed a maturation plan to reduce the number of items on that list. This plan was one of the major risk-reduction activities required by the ATD. The teams also produced a preliminary design of the UDS, which included two aerial vehicles, a ground station, and the necessary support to conduct ground station and in-flight demonstrations of the critical technologies during phase two of the ATD.[136]

In phase two only one industry team, the Boeing Company, was awarded the $110 million contract to continue with the ATD program. Boeing offered the best value to the government in terms of having a design that is effective and affordable, and it developed the best plan for reducing the risks in terms of making the UCAV operational vision a reality. Flight-testing is slated to start towards the end of 2000 and will take place at the National Aeronautics and Space Administration's Dryden Flight Research Center, located at Edwards AFB, California. During the testing Boeing must show all attributes of its UCAV, including aircraft performance and autonomous flight operations to conduct SEAD. Phase two should be complete and UCAVs should fly before the end of 2002.[137]

Summary

The preceding examples show that UAVs and UCAVs have received and continue to receive significant attention in the 1990s. UAVs have been used successfully in two major conflicts, and the Air Force has taken important steps towards determining the feasibility of UCAVs by establishing the UAV Battle Lab and by participating in the UCAV ATD. At the end of phase two of the ATD, the Air Force will have invested $20 million, and DARPA and the Air Force Research Laboratory will have each invested $60 million.[138] Whether or not UCAV operational capability will result from this investment is largely dependent upon the obstacles that UCAVs must overcome. The next section addresses some of the obstacles that UCAVs will face in obtaining operational capability, but first it examines some of the different concepts of operation proposed for UCAVs in the twenty-first century.

Tomorrow's UCAVs

There are different views as to the manner in which UCAVs in the next century should operate. Some concepts are evolutionary, and some are more revolutionary relative to the way manned aircraft operate today. There are some aspects of UCAV concept of operations (CONOPS), however, that are common to most ideas on the subject. For example, most visionaries see SEAD and deep strike as the most probable missions that UCAVs will fly in the near term. There are some who envision UCAVs in an air-to-air or close air support role, but these roles are usually reserved for a long-term, technologically, and operationally mature UCAV system. Most CONOPS incorporate UCAVs into the force structure to complement manned aircraft. There are not many visions that suggest that UCAVs will completely replace manned aircraft in the foreseeable future. UCAV missions could be characterized by the three Ds: dull, dirty, and dangerous. Dull means long-endurance missions that can last for several hours or even days—too long for a pilot to physically endure. Missions such as extended patrols over no-fly zones would fall into this category, or missions that require long-endurance flights to get to the target area and back would also qualify. Dirty missions are those where the threat of biological or chemical contamination is too high to risk sending in a manned aircraft. Dangerous missions for manned aircraft are numerous and growing; and though most combat missions are dangerous, some are riskier than others. Missions on the first days of a war are generally more risky than later missions because the enemy's air defenses will probably be more of a threat early in the conflict. SEAD is one of the most dangerous missions that manned aircraft are used for today, and this would be a prime mission for UCAVs. Using UCAVs for the three Ds would remove a huge burden from the manned combat air forces.[139]

The following are "broad brush" descriptions of some of the CONOPS that UCAV visionaries have developed. The time frame for their implementation covers a range from about two years to 25 years into the twenty-first century. They cover a wide range of originality and technological development as well. Some of the concepts are very similar to the way manned aircraft operate, and some are very different. The concepts that are similar to manned aircraft CONOPS do not require the level of technological maturity and robustness that the more original and creative concepts require. The major concept missing from this group, however, is the concept of having no UCAVs at all. Otherwise, this is a good sampling of the possibilities for UCAVs in the twenty-first century.

Converted Manned Aircraft

The first concept, converting manned aircraft into unmanned strike aircraft, dates back to the early days of aviation; and the idea is still popular almost a century later. Engineers at Lockheed Martin Tactical Systems of Fort Worth, Texas, believe that this evolutionary path to advanced UCAV systems is possible. The company has looked at modifying retired F-16A fighters by removing the cockpit and life support systems and expanding the wingspan from 31 feet to 60 feet. The longer wings can carry more fuel and be fitted with additional pylons to carry more ordnance and sensors.[140] There is also support for converting the Joint Strike Fighter (JSF) into a stealthy, high performance UCAV for about one-half the price of the manned version. Former Air Force Chief of Staff Gen Ronald R. Fogleman publicly expressed support for the idea, and other Pentagon officials proposed that the last few dozen JSFs be built for control by a pilot on the ground or in another aircraft.[141] According to an executive from Missions Technology, Incorporated, "The Air Force could field an unmanned air-to-ground attack jet in two years or less." Missions Technology conducted studies on pulling the pilot from the A-10 Thunderbolt II and converting it into a UCAV to fly low-altitude bombing and strafing missions.[142]

A squadron of manned aircraft converted to UCAVs would operate similarly to a manned fighter squadron. The main difference is that the aircraft would be piloted from remote control sites. Maintenance, logistics, and the employment of the system would all be very similar to the way a manned fighter squadron operates today. However, Maj Jim Shane of the UAV Battle Lab states, "Converting a manned aircraft [to a UCAV] is fine for a short term solution, but in the long term, we need to design from-the-ground-up UCAVs to take advantage of design freedom and better performance."[143]

From-the-Ground-Up UCAVs

DARPA concurs with the concept of from-the-ground-up development of UCAVs in order to take advantage of all of the characteristics of the pure UCAV design. The aircraft would be roughly 40 percent of the size of today's F-16s or F-18s; and to promote stealth qualities, it would carry all weapons internally. There are various conceptual shapes for the airframe; but all would take advantage of stealth designs, perhaps sim-

(Courtesy of *Air Force Magazine*)

A possible design for UCAVs is one similar to the B-2 bomber.

ilar to the B-2 bomber or the F-117 fighter.[144] The design features of a UCAV show great promise for a SEAD role. Without the need to carry a pilot, the smaller, stealthier UCAV would be difficult to detect and shoot down. Such an aircraft could also loiter in an area for extended periods of time—well beyond the limits of a human pilot—until a target is detected. Being close to the point of engagement would allow the UCAV to launch a swift attack. Even if the enemy did get lucky and launch a missile, the UCAV could perform escape maneuvers that a human pilot may not be able to withstand.[145]

In describing DARPA's CONOPS for a UCAV, Lt Col Mike Leahy, deputy program manager of the UCAV ATD program, envisions the UCAV starting out in a storage container at a US base or prepositioned at a base somewhere overseas. When the balloon goes up and the United States becomes involved in a conflict somewhere in the world, the UCAVs are removed from storage. C-17s or C-5s are flown to the storage base, and six UCAVs are loaded on board each C-17 and twelve on each C-5. At least two control stations are also transported to the theater of operations or some remote lo-

(Courtesy of *Air Force Magazine*)

Gull-Wing UCAV with Dimensions—6 Feet Tall, 24 Feet Wide, and 27 Feet Long.

cation. When the transports land at the forward operating base, the UCAVs are taken off the plane and removed from their storage boxes. After the wings are reattached and fuel and weapons are loaded, the UCAVs are ready for combat. By most estimates, this can be done easily within a 72-hour deployment time.[146]

Once in-theater, a team of five operators controls 16 UCAVs. Four operators manage four UCAVs each; and one operator serves as mission commander, overseeing the operation team. The mission commander attends the briefings at the air operations center with his manned aircraft counterparts, and they all receive the air tasking order (ATO). The UCAV mission commander coordinates the same way that a Wild Weasel mission commander would coordinate with other air-to-air and strike aircraft.[147] He then returns to his operators and briefs them on how they are going to carry out the mission. The operators take the missions from the ATO, convert them into detailed mission plans, and program the information into the UCAVs.

The UCAVs take off automatically, fly into the target area, and link up with the strike package. For a SEAD mission, the UCAVs arrive in the target area ahead of the strike package to sweep the area. The amount of autonomy UCAVs are envisioned to have varies; in most concepts of operations, the

(Courtesy of *Military Technology* magazine)

An artist's concept of a UCAV on a single-ship strike mission.

operator does not have to actually pilot the aircraft from the remote control station. The operator monitors the UCAV's mission from the control station and concentrates on aspects of the mission such as mission changes and weapons delivery. The UCAVs escort the strike aircraft on ingress to their targets, and they escort them out on egress.

After landing back at the base of operations, the UCAVs are refueled, reloaded with weapons, and turned around for subsequent missions as needed. When the war ends, the maintenance crews remove the wings from the UCAVs and put the aircraft back into their storage boxes. The aircraft are shipped back to depot, refurbished, and put back into long-term storage until the next conflict. It would also be possible for the UCAVs to self-deploy upwards of 4,000 miles in a single leg and would allow them to be deployed from the continental United States (CONUS) to many distant theaters of operations. Another possibility is to preposition UCAVs in locations such as Diego Garcia, Europe, or Korea—allowing the UCAVs to be deployed even more rapidly.[148] Speed of deployment is important under the DARPA CONOPS because UCAVs are expected to participate during the high-threat, early phases of a campaign. The UCAVs "will penetrate enemy air defenses and provide both preemptive and reactive SEAD and prosecute non-hardened high value targets within the adversary's infrastructure."[149]

AC-17 Mother Ship

Another even more revolutionary idea for UCAVs is the AC-17 "mother ship" concept. Greg Jenkins of the Armament Product Group Manager's Office at Eglin AFB, Florida, envisions the mother ship concept as having potential for tremendous cost savings. Jenkins states that cost savings are not maximized "if a UCAV squadron grows linearly in line with a fighter squadron. You still have a large logistics tail and still have a lot of support equipment and personnel. We have to think non-linearly. That's where the idea of launching from the AC-17 comes from."[150] Jenkins believes that many current concepts of UCAVs resemble fighter squadrons without the pilots; this he says, "is an evolutionary concept." Jenkins and the Armament Product Group Manager's Office promote a more revolutionary concept. "We need to go beyond this mentality and go to the next step, which is the mini-UCAV operated off of a global projection platform. This is how we will really reap the benefits of UCAVs."[151]

The AC-17 mother-ship concept uses a modified C-17 cargo aircraft to carry, launch, control, and recover mini-UCAVs. The AC-17 is modified to carry approximately 18 UCAVs, a control station, fuel, and weapons for the UCAVs, as well as a support crew. The AC-17 is tantamount to an arsenal ship and will dramatically reduce logistics costs.[152]

When the mother ship deploys, it establishes itself in an orbit out of harm's way of any enemy threats. Fighter escort for protection also accompanies it. From its orbit area, the ship launches its mini-UCAVs on their SEAD or strategic attack missions. Each mini-UCAV resembles a cruise missile that is returnable. The vehicles are equipped with ATR and are capable of carrying two 250-pound small smart bombs (SSB) or 48 micromissiles. Each aircraft is launched from the mother ship, flies its mission, and returns to be reloaded onto the mother ship. Once on board the UCAVs are refueled, reloaded with weapons, and sent back out for subsequent missions.[153] The concept is similar to DARPA's CONOPS except that the vehicles are mini-UCAVs and instead of operating from an airfield on the ground, the launch, control, support, and recovery all take place from a "base" in the sky.

StrikeStar 2025

An extremely far-reaching and revolutionary CONOPS for UCAVs is the StrikeStar 2025 concept. The College of

Aerospace Doctrine, Research and Education (CADRE) at Maxwell AFB, Alabama, developed this idea in 1996. The StrikeStar concept emphasizes long loiter times and cost-effectiveness to enable the concept of air occupation—the ability to hold an adversary continuously at risk from lethal or nonlethal effects from the air. StrikeStar would have an 8,000-mile combat radius. Several StrikeStars aloft at any one time "could provide global coverage and operate either as a stand alone system or in conjunction with other forces. Such a capability and the implementation of air occupation would revolutionize warfare."[154]

StrikeStar missions are launched from the CONUS or a forward operating location. StrikeStar performs ground checks, taxi, and takeoff autonomously; and it departs on a spiral climb over the airfield until above the positive controlled airspace. It then proceeds to the target area as programmed unless it receives updated instructions. Once StrikeStar reaches the target area, it delivers weapons in either an autonomous mode or a command directed mode, depending on how it was programmed. If operating in an autonomous mode, it automatically delivers weapons against its assigned target or against targets in its assigned kill box. The kill box is a designated area in which the StrikeStar searches for, detects, and kills targets using its onboard sensors. If operating in the command directed mode, StrikeStar establishes itself in a preprogrammed orbit and awaits targeting instructions from the control center. In both the autonomous and command directed modes, StrikeStar remains on station until it is directed to return or until fuel or weapons expenditures require a return to base. This could be as long as several days. After egress from the area, StrikeStar returns to its home aerodrome's airspace and spirals down for landing.[155] "Because of a StrikeStar's endurance, altitude, and stealth characteristics, it could wait undetected, over a specific area and eliminate targets upon receiving intelligence cues."[156] It is possible that in the twenty-first century a weapon like StrikeStar "could send a lethal or non-lethal message to US enemies and enforce the imposition of our national will through air occupation across the battle space continuum."[157]

These different concepts give a general idea as to the different visions pertaining to the way UCAVs will fight if they become operational. The real question, however, is will they ever become operational?

Issues Facing UCAVs

There are certain driving forces that influence the decisions made by the leaders of America's armed forces, and one of the strongest driving forces is American public opinion. Social, cultural, geopolitical, and economic factors play an important part in all major military decisions. It is likely that as we move into the twenty-first century, these influences will become even stronger and place greater demands on the US armed forces. Two trends that are present today and will probably be even more significant in the future are the US economy driving the armed forces to be more cost-effective and America's sensitivity to the loss of life and treasure in conflict. The Air Force must be sensitive to these factors when planning tomorrow's force structure.

One of the great clichés—we must do more with less—has circulated throughout the Air Force and the rest of the armed forces. The nation has continuously called upon the armed forces to participate in military operations across the conflict spectrum, but at the same time the military budget and force structure have steadily declined during the past decade. Consequently, military leaders have been forced to find less expensive ways of operating without sacrificing effectiveness. This trend will likely continue into the next century.

America has also challenged the armed forces by putting extreme limits on the number of casualties that will be tolerated during military operations. The shootdown of Captain O'Grady over Bosnia is a prime example of America's concern over casualties. Maj Gen Michael Kostelnik, commander of the Air Armament Center at Eglin AFB, Florida, sees this as an important theme in war fighting today. He states that "during the Vietnam War we were losing hundreds of people a week for a long period of time. . . . In general, people accepted this as the price of war. The same was true of Korea and World War II. . . . Today, expectations have changed. We fought Desert Storm in 1991 and lost well under 200 people, but now that we fought that war, we are victims of our own success. Now, society wants us to fight overpowering engagements like Desert Storm and not lose anybody."[158]

Conducting operations along the full conflict spectrum, under the constraints of a tight military budget and with little or no loss of life, is a major challenge that America has given its armed forces going into the next century. To meet these challenges, the Air Force has begun preparing for the twenty-first century, and one major area that has to be addressed is the SEAD capability of the combat air forces. DARPA describes this need as follows:

> The recent trend among our adversaries has been to invest in integrated air defense systems (IADs) rather than aircraft to ensure their own air superiority. The Defense Intelligence Agency (DIA) and service intelligence branches, project those IADS will apply the lessons learned from Desert Storm by becoming more sophisticated, mobile and integrated. With the US and her allies continuing to field new surface based air defenses, proliferation will force the Joint Force to face a Red, Blue and Gray threat array. Large portions of these arrays are mobile and possess improved multi-targeting capability. To counter this asymmetric threat and maintain core competency, the Air Force must maintain an effective and affordable SEAD and precision strike capability.[159]

Members of the DARPA ATD program believe that a successful SEAD UCAV project would take some pressure off the Air Force, which has had no dedicated SEAD platform for seven years. Congress criticized the Air Force for phasing out the F-4G Wild Weasel dedicated SEAD aircraft after the Gulf War without a direct successor in mind. Since the F-4G's retirement, SEAD has been performed in the Air Force by F-16s employing the high-speed antiradiation missile (HARM) and its associated pod. Many believe that the HARM targeting system is a valuable asset but not as comprehensive as the F-4G's avionics suite. There are many who believe that the Air Force should give this critical mission the attention it deserves and a platform of its own.[160] According to Colonel Leahy of DARPA, UCAVs have great potential to meet the Air Force's need for SEAD. Leahy states, "UCAVs offer more flexibility than a cruise missile while still affording no risk to human life, and the potential affordability is significantly greater than operating manned aircraft."[161]

Air Combat Command (ACC) has determined that the combat air forces will be deficient in SEAD capability in the 2015 time frame. According to Maj Fred Zayas of the ACC/DR Advanced Programs Office, "ACC and the combat air forces have a requirement driven by both the threat and affordability with respect to the SEAD mission, and UCAVs have the potential to meet the requirement."[162] In order to have a capability that meets the 2015 SEAD requirements, ACC will have to conduct an analysis of alternatives in the 2005 time frame. This plan gives UCAVs several years to mature as a technology before that time, allowing the new technology to be competitive with other weapon systems. The timing is favorable for UCAVs, and it adds to their potential to meet ACC's needs and aid the combat air forces in waging war in a way that is suitable to the American public.

Technology. A major hurdle that UCAVs face in meeting the needs of the combat air forces is the technological challenge. In the 1970s when the Air Force began to evaluate UCAVs, the technology was not mature enough for them to be an effective weapon system. Since the days of the Ryan BGM-34 class of UCAVs, "UAV technology has gone from being rudimentary to state of the art."[163] Only in recent years has technology matured enough to make UCAVs viable. Technological successes with UAVs and cruise missiles have given UCAV technology a tremendous boost. "[E]merging technologies such as miniaturization of electronics, improvement of sensors, development of reliable and jam-resistant data links, and improvement of navigational accuracy are making it possible to overcome the limitations that [UCAVs] faced in the [1970s]."[164] The challenge now is to determine whether or not UCAV technology is mature enough for the twenty-first century. This section examines some of the enabling technologies that are necessary for UCAVs to fly in the near future.

Airframe. UCAV airframe technology is fairly mature because it is patterned after the same technology in many manned aircraft (B-2, F-117, F-22, JSF). However, there are significant advantages in designing airframes for unmanned aircraft versus manned aircraft. The major impact comes from eliminating the cockpit, resulting in significant reduction in size and weight (about 40 percent smaller than manned aircraft). This reduction has a major impact on the layout of the vehicle in terms of structure, systems, and equipment location. The positioning of internal weapons bays, the location and shape of the air intakes, and the routing of air ducts are much less constrained.[165] The cockpit on manned aircraft is also heavy and bulky, and its position at the front of the plane makes it practically impossible to design an aerodynamically efficient and stable fighter below a certain size.[166] The removal of the cockpit also provides design freedom in optimizing the low-observability, or stealth qualities, of the airframe. "Without a cockpit, the UCAV can be shaped like a manta ray, with all its surfaces sloped as far away from the vertical as possible and with only as much depth as needed to accommodate the engine and weapons."[167]

Weapons. The weapons envisioned to complement most UCAV designs are small smart munitions. UCAV designers anticipate that research into smaller munitions will bear fruit within 10 years, leading to much smaller weapons with as much explosive power as today's 1,000- and 2,000-pound bombs. Using smaller weapons will allow the small UCAV airframe to carry greater numbers of weapons and strike more targets per mission. The smaller weapons will also aid in the stealth qualities of the UCAVs because they will be small enough to be carried internally. Two small smart weapons that are currently under development are the SSB and the low-cost autonomous attack system (LOCAAS).

The driving idea behind the SSB is to produce a weapon that will be as destructive as the standard 2,000 pound, yet so small that three of them would fit into the same space as a single Mk-84 one-ton bomb. This result will be achieved by enhancing the explosive filler in the weapons and by improving their accuracy. The destructiveness of a given weapon varies directly with the weight of the explosive filler, but it varies inversely with the cube of the miss distance. So, there may be more profit in accuracy improvements, making SSB a prime candidate for UCAV employment.[168] Initially the SSB will come with a combination global positioning system (GPS)/inertial navigation guidance system, in a hardened six-foot casing loaded with standard explosive filler. Each weapon will weigh 250 pounds.[169] Later versions are slated to include a more de structive filler and a laser seeker to increase accuracy and physical effects.[170]

Another potential UCAV weapon is LOCAAS, which has been under development since 1990. The LOCAAS program goal is to produce an affordable standoff miniature munition to autonomously search, detect, identify, attack, and destroy SEAD and ground mobile targets of military significance.[171] LOCAAS is to have a standoff capability of greater than 100 kilometers, a search area of 50-square kilometers per weapon, and a cost of $33,000 per unit.[172] Each LOCAAS will be 31 inches long with a wing span of 45 inches, and each will weigh approximately 100 pounds. The weapon will use a laser-radar seeker for ATR, and its multimode warhead will be capable of delivering three different kill mechanisms: a stretching rod for hard armor penetration, an aerostable slug for increased stand off, and fragments for soft target kills.[173] Both the LOCAAS and the SSB would be well suited for use with a UCAV weapon system.

Command and Control. Where the airframe and weapons technology appear to be fairly straightforward and well in hand, the C^2 aspects of UCAV technology present a more difficult challenge. According to General Kostelnik, "The technologies that will make UCAVs capable in the future are not hardware technologies. They are not airfoils, engines, or weapons. We have those technologies, as well as the miniaturization of those technologies at our fingertips. The challenge lies in the software. It's all about connectivity and C^2."[174] UCAVs present a unique challenge with respect to C^2. Colonel Leahy of DARPA admits that "the technology that we don't have as firm a grip on are the areas of C^2 and the man-machine interface. These areas are unique to unmanned systems and they are crucial with UCAVs because deadly force has to be authorized."[175] The risk of temporary, partial, or total interruption in the direct links between the human operator and the UCAV's onboard sensors must be carefully evaluated when planning a UCAV mission. The resulting loss of situational awareness would make it difficult to complete the mission successfully. If the operator could not positively identify the target via the UCAV sensors, he could not authorize weapons release.[176] This situation could endanger the UCAV, and, more importantly, it could endanger an entire strike package that depends on the UCAV for SEAD. "[T]he possibility of natural or man-made electromagnetic interference or novel forms of information warfare resulting in [a catastrophe] cannot be ruled out that easily."[177] Developing secure, over-the-horizon, antijam data links is crucial to the future of UCAVs.

There are also concerns about the amount of bandwidth available for UCAV operations. Sending information such as coordinates or text does not require inordinate amounts of bandwidth; but when pictures of targets must be sent, the amount of data that travels through the bandwidth expands significantly. If video is sent, the amount of data becomes enormous; and if this amount is multiplied by several UCAVs, potential problems arise. Many users in the area of operations compete for the limited bandwidth; and if the data-link system gets overloaded, it may shut down altogether. Currently there is not a data link with a broad enough bandwidth to handle this kind of volume.[178] Though DARPA studies have shown that data-link and communications capabilities in the 2010 time frame will overcome these bandwidth and communications problems, these solutions must be demonstrated and proven before UCAVs can become operational.

Autonomy. Inversely related to C^2 is autonomy; the more autonomous the system, the less C^2 required. There is a spectrum for the levels of autonomy being considered for UCAVs. At one end of the spectrum is full autonomy, where the UCAV would fly to the target area via programmed way points, turn on its ATR system, and kill any target it recognizes. At the other end of the spectrum is a situation where the UCAV flies to the target area, detects a target, and sends the image back to the control station. The operator then verifies the target and sends back authorization before the UCAV delivers weapons.

The level of autonomy that is used depends greatly on the level of technological maturity. Artificial intelligence is a challenge, particularly in a fully autonomous situation where there is no man in the loop. The UCAV has to display some level of human judgement. According to Major Zayas, "We currently have a man doing the job that we are considering UCAVs for. That man makes many decisions regarding targets, engagements, and threats. The pilots that do this now have years of experience and know how to make those judgements. Trying to mimic that in a machine is a big challenge."[179] Though technologies such as ATR are becoming increasingly mature, the level of artificial intelligence that would allow UCAVs to integrate and operate with manned aircraft and react to the fog and friction of war is not yet available.

Technology Integration. From DARPA's point of view, the real technological challenge is to take present technologies, mature them, and integrate them into a functional UCAV system.[180] This integration has to be done practically so that technology is usable for the operational air force. It has to be done safely so that the UCAV can integrate with other combat operations, and it has to be done effectively in terms of engaging the target and completing the mission within the C^2 structure. Finally, it has to be done affordably, which is the topic of the next section.[181]

Cost-Effectiveness

The issue of cost-effectiveness will be influenced by everything from technology to training to combat operations. In each of these areas, trade-offs have to be made and cost-effectiveness solutions have to be proven. It not only has to be proven that the solution will be affordable but also that the solution will adequately accomplish the stated mission. This section discusses system costs and O&S costs to determine where the potential savings will be made.

System Costs. DARPA estimates that a UCAV will be approximately one-third the cost of a JSF, and the JSF itself is intended to be inexpensive compared to the cost of earlier fighters. This estimate would put the cost of a UCAV at about $11 million as measured in 1999 dollars.[182] UCAVs are inexpensive because they do not need the life support systems, instruments, or escape systems required by manned aircraft. They will be smaller and will require fewer materials, which will also add to the savings.[183] Unlike a manned aircraft, UCAVs are not likely to fly except in combat and for occasional tests and exercises; therefore, each one can be designed for minimal flights—making it possible to build the aircraft less expensively.[184]

Operation and Support Costs. The biggest potential savings, however, will come from reduced operation and support costs. On paper, a UCAV squadron's lifetime O&S costs will be 25 percent of the lifetime O&S costs of an equivalent F-16 HARM targeting system squadron.[185] A large portion of the savings will come from the idea of leaving UCAVs in dormant storage for most of their service lives while they await the call to action.[186] A few aircraft would remain out of storage to enable personnel to practice loading weapons and other maintenance type functions. There would also be one active squadron that could support training exercises like Red Flag, dramatically reducing the amount of training hours and resulting in significant savings.

Operators are expected to maintain proficiency by practicing in a virtual environment. They would train using the same equipment that they would use in a real conflict. They would experience the same visual and aural cues that they would experience on an actual mission. Training this way avoids operating costs such as fuel, spare parts, and maintenance.[187] Another possibility is the training of reservists to maintain UCAVs. This training would result in the ability to activate maintenance personnel when the UCAVs are activated. For the same reason, many of the operators could be reservists. A core group of operators and maintainers would remain on active duty to respond to short-notice contingencies and training requirements.[188]

O&S costs would be reduced further by a change in crew ratios. The USAF maintains a pilot-to-fighter ratio of about 1.3 to 1. With UCAVs the ratio will be reversed. One operator will manage several UCAVs at once. The operator to UCAV ratio will be between four and six to one and will be possible because of the high degree of onboard autonomy that UCAVs are expected to possess. The UCAVs are envisioned to take off, fly to the target, and return to base on their own.[189] The operator, under normal circumstances, would only be involved in authorizing weapons delivery. With the UCAV accomplishing some of the basic flying tasks on its own, the operator is free to concentrate on the cognitive aspects of the mission such as planning the mission execution.[190]

There are many promising, but mostly notional and unproven, concepts that allow UCAVs to be more cost-effective. Some of the technologies that are required to transform these concepts into reality are not yet mature enough to be integrated into the overall UCAV concept, and the actual cost of these technologies has not yet been finalized. For example, the technology required for UCAVs to be autonomous will be new and expensive. Col Paul Geier, chief of Combat Applications at the UAV Battle Lab, agrees that a more autonomous system will save dollars through reduced manpower requirements; but he also believes that the technology required to achieve this level of autonomy will be costly. There are trade-offs.[191] The Predator UAV requires as many as six personnel to conduct a single mission, which is a significantly less complex mission than that envisioned for UCAVs. Going from a six to one ratio to a one to six ratio is a change by a factor of 36 in personnel required. Though this is a worst case estimate, it suggests that there are a lot of technological improvements needed before UCAVs can realize the cost-effectiveness vision set forth for them. Colonel Geier summarized the cost-effectiveness issue by stating that "UCAVs have to be cheap, but if they're so cheap that they cannot accomplish the mission, they are a waste of resources. If they're very expensive, but do the mission extremely well, however, they run into competition with other systems that do the job as well, for less money."[192] This competition with other systems is an important issue, and it is also the next subject for discussion.

Competing Weapons Systems

Throughout the history of unmanned aviation, UAVs and UCAVs have had to compete with manned aircraft and cruise missiles for funding, operational use, and esteem in the aviation world. Today's UCAVs are faced with the same competition plus an emerging competitor—space-based systems. This section will compare and contrast UCAVs against the three types of competing systems.

Manned Aircraft. Under most concepts UCAVs do not replace manned aircraft, but they do complement manned aircraft by flying those missions most appropriately suited for unmanned platforms—dull, dirty, and dangerous. On paper, as previously discussed, UCAVs offer a tremendous cost savings over equivalent manned systems. UCAV technology also promises to be as effective as manned aircraft, particularly in the SEAD role. The first mission that UCAVs fly will most likely be SEAD, which is one of the most dangerous combat missions flown. Without a human on board, UCAVs will be able to take some risks that otherwise would not be prudent. This point leads to what is probably the most obvious benefit that UCAVs have over manned aircraft—UCAVs can guarantee that no friendly lives will be risked to accomplish the mission.

Although the UCAV is potentially a high payoff system, it is also a higher risk solution than manned aircraft. A UCAV is a higher risk solution to the requirements of the combat air forces because it is a new technology, and leaders and planners have little or no experience with UCAVs. They do have confidence, however, in manned systems such as JSF and F-22. The Air Force has a long history of developing, procuring, and modifying manned aircraft; and there is more corporate knowledge in dealing with the more mature technology. The other risk in employing UCAVs is integrating them into the force structure. They may have to operate with manned aircraft, and this integration presents problems that the Air Force does not currently have. One such problem is airspace management and deconfliction. This area is key to successful operations in civil and military environments. UCAVs "must operate in diverse airspace environments, so appropriate approaches to airspace deconfliction are essential."[193] In controlled airspace, the Federal Aviation Administration has traditionally operated under the rule of "see and avoid," but it is not possible for UCAVs to follow this rule in its traditional sense. Modifications to the rules of airspace management will have to be made. Deconfliction in a combat environment is another area that UCAVs will have to contend with. This is a difficult task when planners are only concerned with manned aircraft; the task of deconflicting a strike package made up of both manned and unmanned aircraft will be even more daunting. A final advantage that manned aircraft have over unmanned aircraft is greater flexibility. It is unlikely that technology will give rise to a computer that duplicates the human brain, all five human senses, judgement, and human intuition in the near future. Having a man in the loop can make up for some of this loss of human qualities and flexibility, but it will not be the same as having a man on the scene.

Cruise Missiles. Another type of weapon system that the UCAV will have to compete with is one of its early ancestors—the cruise missile. They are very similar in many ways, but the major technical difference is that cruise missiles do not come back when the mission is complete. The fact that UCAVs are reusable gives them an advantage in terms of cost-effectiveness. With UCAVs employing small, inexpensive, smart weapons, they could destroy targets at a fraction of the cost of weapon systems such as the $1 million Tomahawk land attack cruise missile (TLAM). Each time a TLAM destroys a target, that weapon system—including its engine, airframe and sensor suite—is lost. With a UCAV the only part of the system that does not return for subsequent use is the munition.

UCAVs also promise to be more flexible than cruise missiles. UCAVs have the capability to accept mission changes, abort missions without self-destructing, and strike relocatable targets. Having a man in the loop affords UCAVs a great deal of flexibility. Colonel Leahy points out that "cruise missiles are launched from a long way out. UCAVs operate from much closer because of their reduced signature. This will allow for better target acquisition, and it will compress the time between the moment the decision is made to destroy the target and the moment the target is actually destroyed. It will allow UCAVs to compress the OODA loop more effectively than cruise missiles."[194]

There is a class of standoff weapons that includes the joint standoff weapon (JSOW) and the joint air-to-surface standoff missile (JASSM) which can be launched from manned platforms. These weapons are very capable and are competitive with UCAVs; but because these systems must be launched from manned platforms, all of the advantages that UCAVs will potentially have over manned aircraft will also be advantages over these systems. These advantages include cost-effectiveness and minimal risk to pilots; and if the aircraft stands off to minimize the risk to the aircrew, the cruise missiles lose their flexibility and perform similarly to other long-range missile systems like TLAM.

There are, however, some advantages that cruise missiles have over UCAVs. The major advantage is that the US armed forces are familiar with cruise missiles. The technology has been around for a long time, and the risk in investing in further cruise missile technology is low compared to UCAVs. They have been successfully employed in combat, and they have been integrated with manned assets successfully. Cruise missiles do not have the airspace concerns that UCAVs have because planners have learned how to deconflict airspace for cruise missile use. C^2 is not nearly the issue with cruise missiles as it is with UCAVs. Cruise missiles are fully autonomous, and their C^2 and bandwidth requirements are minimal. In many ways UCAVs are a cross between cruise missiles and manned aircraft. The key is to find a way to allow UCAVs to capitalize on the advantages of both types of systems and minimize the disadvantages.

Space-Based Systems. A third potential competitor for UCAVs is the space-based system. There are many concepts within the DOD's research and development circles that employ space-based weapons systems capable of delivering precision force against ground-based targets. These systems will have the potential to project power to any point on the earth with minimal sensor to shooter delay, and they will provide decision makers a near continuous coverage of all global hot spots.[195] The greatest advantage that space-based systems will have over UCAVs is that it will take time to mobilize and deploy UCAVs to the theater of operations. Space-based systems will be in place continuously and will strike at a moment's notice.

The disadvantages of space-based systems are mostly associated with costs and socio-political concerns. Research and development is expensive and so are procurement and operating costs. The costs associated with transporting the space vehicle from the earth's surface to an earth orbit, maintaining it, and then transporting it back to the surface will be significant. "Another significant space-based limitation is the criticality of the vehicle's position or orbit. Space-based systems cannot currently loiter over a target since orbital mechanics require constant movement around the earth. Therefore, a space-based system needs multiple vehicles to provide constant coverage as well as the ability to position a vehicle when and where needed."[196] This would add significantly to the overall system cost. One must also consider the social and political implications of militarizing space. "Establishing space dominance will be costly and threatening to an increasingly interdependent international community. Placing an offensive-capable platform in space that continuously holds any nation or group of individuals at risk will undoubtedly be perceived as a direct threat to friendly or enemy nations."[197]

Service Cooperation

Throughout the evolution of UCAVs, the services—particularly the Navy and the Air Force—have worked simultaneously on similar projects without combining their efforts and resources. Both efforts often failed. Congress is also concerned with this issue. A 1997 congressional report stated that, "Congress has for many years been concerned with the problems of waste and duplication of effort in the defense budget. It has proven administratively difficult to persuade or compel the services to merge their research and development and procurement efforts."[198]

The Navy has shown some interest in UCAVs in recent years. According to George Palfalvy, an operations research analyst at the Naval Air Warfare Center's Weapons Division, "UCAVs are the natural product of three converging trends: (1) the need for pilots to concentrate on higher order tasks than simply flying an aircraft, (2) the demand for more capable cruise missiles, and (3) the growing ability to build and operate very sophisticated unmanned aerial vehicles."[199] Early in 1998 several industry teams conducted studies that looked at different ideas for an unmanned naval strike aircraft, with emphasis on those requirements unique to launch and recovery from naval platforms. The teams looked at three types of naval UCAVs. Two were designed to operate from surface combatant vessels, and one was designed for launch from a submarine.[200] Though the studies produced several different designs, the Navy has yet to pursue UCAV technology further.

The UCAV ATD is a joint DARPA/Air Force effort, and the Navy is not involved. According to Colonel Leahy, "DARPA co-

(Courtesy of *Air Force Magazine*)

Industry teams looked at UCAVs launched from submarines *(left)* and surface ships *(right)* at the request of the US Navy.

ordinates with the Navy counterparts and most of what we are doing will be directly applicable to them. We will continue to pull them into the program as we move along, but it was difficult enough to get DARPA and the Air Force to work together. A third party would have gummed up the works more."[201] Larry Birckelbaw, the UCAV ATD program manager, observes "there is the expectation that if it proceeds to a full-fledged developmental effort, before full production, there is an interest in making it a joint program."[202] The UCAV ATD is seen as "taking the lead in exploring the tactical utility of UCAVs, and the US Navy will probably wait for the results of the land-based demonstration before deciding whether unmanned strike aircraft have a place on the decks or in the tubes of a future seaborne fighting force."[203] It appears in this case that there is no duplication of effort with respect to UCAVs. The Air Force/DARPA team leads the program; and if the concept proves feasible, the Navy seems prepared to join the effort. Until that time, however, the program will remain under the current management, which leads into the next issue for discussion.

Management

Historically DOD has had a poor track record when it comes to managing UAV programs, and that trend continues today. Since 1979 the DOD has been involved in eight UAV programs. Of those, four have been terminated, two remain in development, and only two have been fielded as operational systems. DOD has spent more than $2 billion on the development or procurement of these eight programs during the last 18 years.[204] The latest UAV casualty was DarkStar, the high-flying, stealthy reconnaissance UAV, which was canceled early in 1999. These numbers indicate potential problems in the management of UAV programs, and it remains to be seen whether the ongoing UCAV ATD will meet the same fate. According to DARPA and ACC, however, they have learned from the past mistakes of UAV program managers, and they have taken steps to prevent the UCAV ATD from meeting a fate similar to that of many UAV programs in the last 20 years.

The purpose of the UCAV ATD program, which started in 1997, is to evaluate the available technologies, combine them into an operational concept, and determine if the resulting system could effectively and affordably address the SEAD mission.[205] According to Colonel Leahy, the program could have been an advanced concept technology demonstration (ACTD), but that type of program was not appropriate for the pursuit of UCAVs.[206] In an ACTD, once the technology is developed, it goes directly into the field. Improvements needed to make the technology more operationally useful are difficult to make once the system goes into the field.[207] Predator was an ACTD and ran into this very problem. Its most notable deficiency was its lack of all-weather capability, which severely reduced its effectiveness in Bosnia.[208]

Contrary to the ACTD, an ATD allows the technology and the operational aspects of the system to mature before it goes into the field. The technology does not have time to develop into full capability when it has to go into the field right away, and the technology's path to operational development is set prematurely because the full technological potential is unknown. The ATD process allows the technology to develop and to have its full capability explored.[209] The UCAV ATD, however, is not a traditional ATD. DARPA has incorporated a significant amount of user involvement into the program as well as a great deal of commitment from the acquisition community and the laboratories. According to DARPA's Mike Leahy, "It is unusual to get this amount of buy in by all of the parties."[210] He goes on to say, "This program is very focused on producing something the user wants. ACC has been in on this program from the beginning. They helped write the phase I and II solicitations and they have been on the evaluation boards picking the winners."[211]

Major Zayas, from ACC's Advanced Program Office, agrees with Colonel Leahy: "The user, ACC, is deeply involved in the process. This is rare because historically, the user doesn't get involved in the development of new technologies at such an early stage."[212] Boeing UCAV program manager Rich Alldredge believes the SEAD project has skipped years of potential dead ends because it is "the best example so far of a technology demonstration program working with the ultimate user." ACC, he said, has provided "invaluable insights" in the concept development studies and steered the project toward what would be most useful. In other projects where the coordination has not been as tight, "we've missed the mark."[213]

To help the program "hit the mark," ACC provided operational guidance to help develop the concept more fully. Examples of this guidance included the following:

— ACC provided information on conducting SEAD operations, both reactive and preemptive and in support of a strike package.
— ACC designed the air campaign models that DARPA used to verify and improve their UCAV designs.
— Guidance with regards to limiting a logistics footprint was provided.
— DARPA was provided with information on air base operations and all of the tasks that are necessary for flying in and out of an airfield.
— ACC provided guidance as to what maintainers would face with respect to UCAVs to aid in the development of their maintenance and supportability concepts.
— Employment concepts were provided to help develop plans for moving from stateside bases to the area of operations.

— Guidance was provided on control console operations and on tasking console operators.[214]

The managers of the UCAV ATD are trying to avoid mistakes of the past. Programs such as Predator, Global Hawk, and the recently canceled DarkStar were all promising technologies; but because a lot of operational aspects were not addressed early, the systems did not meet the requirements of the user. By getting involved early in the UCAV ATD, ACC is ensuring that the operational implications are being addressed before it is too late. What they are trying to avoid is a system that looks good on paper, but does not address all of the operational aspects. They want to prevent the $10 million project from becoming an $80 million project in order to meet operational needs. This approach to the ATD allows the operational aspects of the system to mature with the technological aspects. When it is time for ACC to evaluate the system for purchase, they will know whether it is operationally sound or not.

Treaties and Agreements

If UCAVs do become operational, there are some arms control agreements and treaties that could potentially limit or prohibit their use. Although no arms control agreements limit UCAVs directly, the Intermediate-Range Nuclear Forces (INF) Treaty and the Strategic Arms Reduction Treaty (START) have the potential to limit them indirectly. "Strict reading of the INF definition of 'cruise missiles,' that is, 'an unmanned, self-propelled weapon delivery vehicle that sustains flight through the use of aerodynamic lift over most of its flight path,'" would bring UCAVs under control of the treaty.[215] This means that as specific UCAV designs are determined, "they will require DOD Compliance Review Group analysis early in the program for a case-by-case determination of prohibited or permitted fielding under START and/or INF."[216] According to the USAF Scientific Advisory Board, however, "the treaty provisions should not preclude or limit [UCAV] technology development, for there is precedent for excluding [UCAVs], and it is our belief that other [UCAVs] could be excluded as well."[217]

Political Support

Another potential inhibitor to UCAV development is Congress; in general, Capitol Hill has supported UAV development with the necessary funding. According to the Congressional Research Service (CRS), Congress has been more aggressive towards funding UAVs than it is towards many other weapons systems: "The role of Congress in encouraging the acquisition of UAVs differs from the usual pattern in which a service initiates new technologies and seeks congressional authorization and appropriations. UAVs have never had pervasive and determined support by any service, but armed services and appropriations committees in both houses have sought to encourage the procurement of UAVs because of their comparatively low cost and their utility to military operations and to do so in a way that would avoid unnecessary duplication among any programs that might be initiated."[218]

Louis Rodrigues, director of Defense Acquisition Issues for the Government Accounting Office, stated, "We are very much in favor of Unmanned Aerial Vehicles. We think they offer tremendous potential. . . . What we are dealing with is how do you go about getting what we need so that it works and so that we can do that in a cost-effective manner."[219]

Though the support for UAVs among civilian leadership appears to be strong, there is a concern that UAVs have not been fielded; and there is frustration over the apparent slow progress of UAV programs. In his opening statement to the Military Procurement Subcommittee in 1997, Rep. Duncan Hunter of California, chairman of the Military Procurement Subcommittee, expressed his concern over the procurement of UAVs:

> It seems that over the last 20 years, we have spent billion of dollars developing a variety of [UAV] platforms, but for some unknown reason—and maybe we will find out about that reason today—there are precious few assets in the inventory.
>
> I do not know whether it has been changes in requirements, changes in management organizations, changes in acquisition philosophies, changes in resource allocation or a combination of these factors that has plagued the fielding of operationally effective UAVs over the years. All I know is that we put a man on the moon within a decade from the time we undertook this endeavor until it actually happened; yet as my colleagues are about to find out from our CRS and General Accounting Office (GAO) briefers, we have not been able to put more than a few UAVs into the hands of the warfighters after nearly two decades of trying to do so.[220]

This is the same concern that Congress had in the 1970s regarding UAVs. They wanted to know why there had been no return on the investment in the technology. Rep. Owen Pickett of Virginia, a member of the Military Research and Development Subcommittee, asked, "From the standpoint of our nation's ability to utilize or to capitalize on the benefits to be derived from these unmanned vehicles, is there a deficiency in our technology, our manufacturing ability, or our understanding of what is required to make these things work?"[221] These types of pointed questions are indicators of the interest Congress has in unmanned aviation technology, but it is likely that continued failures, such as the recent cancellation of the DarkStar program, will begin to erode their confidence in and support for UAV programs. Lack of support from the Pentagon could also erode this support, and there is a perception among many congressional members that Congress supports UAVs more than the Pentagon does. One Senate Armed Services Committee staffer stated, "[I]f there is no Pentagon support, there is little Congress can do" to make more UAVs operational.[222] In his view the reluctance to procure UCAVs does not exist on Capitol Hill; the reluctance is coming from the leadership at the Pentagon.

Pro-Pilot Bias

There are many who believe that Air Force leaders are reluctant to support UCAV and UAV technology because of a pro-pilot bias, sometimes called the "white scarf syndrome." The idea behind this syndrome, as described by Carl Builder, is that when Air Force leaders are faced with a choice between their preferred means (the airplane) and embracing alternative means (UCAVs, UAVs, missiles, etc.) to conduct airpower, they will choose the airplane regardless of other factors. This choice reveals that their true affection is not for combat effectiveness, but for the airplane.[223] According to this line of reasoning, UCAVs have not and probably will not achieve significant operational capability because Air Force leaders are more focused on the preservation of manned aircraft than on what is best for the Air Force.

There are several examples, however, that refute Builder's argument. Though there may be some in the Air Force that suffer from the white scarf syndrome, there have been numerous Air Force leaders who have looked to means other than manned aircraft to enhance airpower. In 1996 General Fogleman, former Air Force chief of staff, stated, "The bottom line is that, on my watch, the Air Force will embrace UAVs and work to fully exploit their potential."[224] His leadership was essential to the progress that UAVs have made in the USAF during the 1990s. Another former chief of staff, Gen Curtis E. LeMay, who was known as a champion of the manned bomber, wrote this about guided missiles: "The long-range future of the [Army Air Forces (AAF)] lies in the field of guided missiles. Atomic propulsion may not be usable in manned aircraft in the near future, nor can accurate placement of atomic warheads be done without sacrifice of the crews. In acceleration, temperature, endurance, multiplicity of functions, courage, and many other pilot requirements, we are reaching human limits. Machines have greater endurance, will stand more severe ambient conditions, will perform more functions accurately, will dive into targets without hesitation. The AAF must go to guided missiles for the initial heavy casualty phases of future wars."[225] These statements show General LeMay's willingness to embrace an alternative form of airpower in lieu of manned aircraft to increase combat effectiveness.

One of the most successful unmanned platforms in Air Force history is the intercontinental ballistic missile (ICBM); and the man that brought it into the Air Force, Gen Bernard A. Schriever, is a prime example that refutes the white scarf syndrome. During his 30-year military career, General Schriever served as a World War II bomber pilot and held many high-level positions in the Air Force. In 1954 he became commander of the Air Force's Western Development Division (WDD) in Inglewood, California.[226] As WDD commander he directed the nation's highest priority program—the development of an ICBM. He was responsible for pushing forward research and development as well as providing the launching sites and equipment, tracking facilities, and ground support equipment necessary to these missiles.[227] WDD produced the Thor, Atlas, Titan, and Minuteman missiles. General Schriever also assumed responsibility for Weapon System 117L (WS117L), the first Air Force satellite program. WS117L led to missile warning, communications, meteorological, and other space-based capabilities.[228]

General Schriever said his group accepted that they were taking risks. He knew, however, that if an ICBM long-range capability and satellite reconnaissance system were not developed, there would be a major instability in the strategic balance between the United States and the USSR. According to Maj Gen Richard D. Curtin, an integral member of Schriever's team, "As opposed to existing airplanes, we didn't know if these things would work." He added that a "few naysayers believed that precious resources were better spent on proven weapons systems."[229] Despite the opposition, however, General Schriever and his team remained immune to the white scarf syndrome and brought the ICBM on-line for the Air Force.

A final example of an Air Force leader who breaks the mold of Builder's concept is General Kostelnik, commander of the Air Armament Center at Eglin. He is a test pilot with experience in several different types of aircraft and admits that he "would have every interest to say that we should strictly rely on manned aircraft." General Kostelnik believes that there will be a role for manned aircraft like the F-22 and the JSF in the foreseeable future, but he recognizes an upcoming need for a weapon like UCAV. "I see from a technology push and a requirement pull that there is a mission for UCAV. We are becoming less willing to risk the lives of our pilots, and we need some high fidelity, accurate, dependable capability that will go in and do the job." The general sees cruise missiles, such as TLAMs and air launched cruise missiles as an option; but he believes UCAVs are a better choice for the future. He goes on to say that "if we had a UCAV that could carry small GPS-guided smart munitions with a man-in-the-loop . . . it could go in where we are unwilling to risk a man and do the same job that cruise missiles are doing, but with much higher fidelity and affordability."[230]

These are just a few examples of high-ranking Air Force leaders who recognized the value of alternative means of conducting airpower. Though this is only a small sample of leaders that refute the white scarf syndrome, history shows that the USAF has led the way in developing and employing unmanned airpower throughout its existence. It developed ICBMs in the 1950s, employed UAVs in the 1960s, experimented with UCAVs in the 1970s; and today the Air Force is leading the way in UAV, UCAV, and space technology.

If there is a reluctance to embrace alternative means of employing airpower among Air Force leadership, one possible explanation that must be considered is the leadership's concern about the effectiveness of unmanned aviation technology.

Instead of "an affection for airplanes," perhaps the Air Force suffers from "an aversion to risk." There is risk and uncertainty involved in embracing new technology. When dollars have to be diverted away from a proven technology to fund an unproven one, caution would seem to be a prudent approach. The Air Force was built on the faith that airpower pioneers had in manned airplanes, and America's combat pilots from past wars have staked their lives on faith in the machines they flew. The Air Force in general has a long history of placing its faith in aircraft, and it appears to be difficult for Air Force leaders to risk placing that faith in some other form of airpower.

The Air Force's aversion to risk and uncertainty is illustrated by the words of a memorandum presented to Gen Hoyt S. Vandenberg, former Air Force chief of staff: "Pilotless Bombers [guided missiles] will replace Piloted Bombardment Units when operationally proven."[231] This attitude shows that the Air Force was unwilling to take a risk on an unproven technology. This is the same attitude that Undersecretary of the Air Force Plummer exhibited toward UAVs in the mid-1970s. He was quoted earlier in this study as saying, "We are going to have to be convinced of the operational utility of a system before we initiate a full-scale development program, even if the program cost is projected to be small."[232] This statement indicates that risks and uncertainties would not be tolerated. When asked whether he had witnessed the white scarf syndrome biasing attitudes about UCAVs at headquarters ACC, Major Zayas replied as follows:

> There is a tendency for pilots to be skeptical until the technology is proven to them, and you can't blame them, especially when you're dealing with a SEAD mission. I'm trying to convince a pilot or a commander to trust this machine that is going to takeoff without a human in it and is going to protect his strike package, allowing it to get through to its targets. They are more skeptical because they know about the fog and friction of war. They know that it is difficult to replace the person in the cockpit, who possesses the necessary judgment to deal with things happening in real time. There is a lot of skepticism that the human can be duplicated. The pilots are also skeptical because they know how difficult the job is, and it is tough to convince them that a machine could do it as well. They often have the attitude that, "I can't get the computer on my desk to work everyday, how can I trust my life and my airmen's lives to one."[233]

The reluctance seems to boil down to risk and uncertainty, rather than the love of manned airplanes. As Stephen P. Rosen concludes in *Winning the Next War:*

> The management of military technological innovation in the modern US military appears to have been dominated by the problem of uncertainties about the enemy and the costs and benefits of new technologies. The evidence from the periods before and after World War II does not show that the civilian scientific community had any inherent advantages over the military in the management of this uncertainty. Scientists made numerous, invaluable contributions to military technology. Military organizations made numerous errors in choosing which technological avenues to pursue, and military men clashed with scientists. But there is no evidence that scientists had to intervene so that the military would make use of available technologies that were being neglected because of bureaucratic pathologies.[234]

If UCAVs are proven to be technologically and operationally sound, and the risk and uncertainty surrounding them can be managed, it is likely that some of the reluctance that some mistakenly identify as the white scarf syndrome will disappear. UAVs, for example, have been proven to be operationally worthy; and they have been embraced as a viable airpower capability. This capability is illustrated by recent events in Bosnia where a Predator operator safely landed his UAV after it suffered engine failure. The operator was awarded the Air Force Aerial Achievement Medal for his efforts.[235] This is an example of the value that the Air Force puts into "alternative means" of achieving airpower. Risk and uncertainty must be managed in order for UCAVs to have the type of success that programs such as ICBMs and UAVs have had, but managing this risk and uncertainty will be a significant challenge.

Conclusions

The history of unmanned flight began over 2,000 years ago when a pioneering Chinese aviator piloted the first kite over a breezy patch of Chinese landscape. The kite sparked the imagination of inventors around the world for hundreds of years, and eventually led to the invention that would change aviation and the world forever—the airplane. Though unmanned aircraft held an inferior position to manned aircraft after the airplane's invention in 1903, unmanned aircraft nevertheless continued to evolve.

UCAVs Yesterday

The Flying Bombs of World War I were among the first unmanned aircraft designed to deliver weapons; though very similar to cruise missiles, they were the earliest ancestors of the modern UCAVs. The advent of radio control during the interwar years allowed the US armed forces to experiment with the conversion of manned aircraft into unmanned strike weapons during World War II. Most of these efforts were unsuccessful, but this did not stop UCAV evolution. During the cold war, remotely piloted aerial target drones were converted into reconnaissance UAVs to peer over the Iron Curtain. It was not until the Vietnam War, however, that UAVs got the opportunity to prove their operational worth. The Ryan 147 was the workhorse UAV during the war in Southeast Asia, and its success led to the expanded role of UAVs which led to the development of the UCAV. The first modern UCAVs were modified Ryan 147s; they were designated as the BGM-34A. Three variants of the BGM-34s were developed, but none ever achieved operational capability. Shortly after the cancellation of the BGM-34 program, all Air Force UAV programs were put on hold for over a decade. The evolution of UCAVs from the flying bombs in World War I through the cancellation of the BGM-34s in the late 1970s was riddled with obstacles. The obstacles included technological deficiencies, managerial impediments, political timidity, lack of service cooperation, a pro-pilot bias, competing weapon systems, and cost-effectiveness. These obstacles added to the UCAV's failure to achieve operational capability and stifled the Air Force's pursuit of UAV technology in general. Throughout the 1980s, Air Force UAV activity was essentially nonexistent.

UCAVs Today

The Gulf War in 1991 renewed the Air Force's interest in UAV technology. Though it did not employ any UAVs in the war, the Air Force noted the contributions of UAVs employed by the other services. USAF UAVs were used over Bosnia and the Balkans beginning in 1995. In 1997 the UAV Battle Lab was established at Eglin AFB to explore UAV capability and technology and report its findings to the corporate Air Force to aid in decisions regarding UAVs. The UCAV ATD program was initiated in 1997. This joint DARPA/Air Force program has the goal of demonstrating the technical feasibility of a UCAV system that will effectively and affordably prosecute twenty-first century SEAD/strike missions.

With the Air Force aggressively pursuing UCAV capability once again, two questions must be asked: What are the obstacles that UCAVs may face in achieving meaningful operational status in the USAF, and can these obstacles be overcome? To answer these questions, this study revisited the obstacles that the UCAV faced throughout its evolution as the basis for determining possible obstacles facing the Air Force's current quest for UCAV capability.

UCAVs Tomorrow

Aviation technology has advanced significantly in the last 20 years, and UCAV research has profited from these advancements. Much of the airframe and weapon technology that will be used for UCAVs is similar to the technology that has been or is being developed for manned aircraft. Miniaturization technology is at hand also, but the major challenge regarding all of these close-at-hand technologies is making them affordable. Some key technologies that are not quite so close at hand are C^2 technologies (e.g., improved data-link and bandwidth capability) and artificial intelligence technologies (e.g., ATR). Though these technologies are available to some extent, they have not yet matured to the level required by operational UCAVs. The greatest technological challenge, however, will consist of integrating all of the matured technologies into a UCAV operational system.

Although there is some risk involved in relying on specific technologies to mature, past history indicates that this is an acceptable risk. According to General Kostelnik, commander of the Air Armament Center: "One of the biggest flaws in our thinking is that we are thinking about tackling a problem ten years from now with the capability that we have today. We have to quit thinking about ourselves in the way we are today and the way we have been in our past. We have to look back at our past, see how much things have changed and project a different future."[236]

Computer technology is an excellent illustration of General Kostelnik's point. "In 1975, an IBM mainframe computer that could perform 10,000,000 instructions per second cost around $10,000,000. In 1995 (only 20 years later), a computer video game capable of 500,000,000 instructions per second was available for approximately $500."[237] If this is any indication of how much technology will mature in the next 10 or 20 years, the chance that the necessary technology for UCAVs will become sufficiently mature is quite promising.

Advanced UAV technology is currently very costly. The recently failed DarkStar program is a prime example. The DarkStar ACTD was supposed to develop a high-altitude reconnaissance UAV for approximately $10 million; unfortunately, $10 million was not enough to create the stealthy UAV. According to J. A. Blackwell, president of Lockheed Martin, "Because the price was $10 million, we couldn't put in the robustness that was needed . . . [to make DarkStar] a production type vehicle."[238] With UCAVs being much more complex than the reconnaissance UAV DarkStar, it is likely that a UCAV purchased today would cost more than the projected $11 million. This injects some risk into the pursuit of UCAVs; but if the trends in technological advancement continue, the risk is worth taking.

There are also some concepts that will significantly decrease O&S costs that must be validated. What some call the "wooden round" concept, which puts UCAVs in storage until they are needed, must be proven viable. This concept would result in significant savings because UCAVs would only be operated during times of conflict. Another cost-saving idea is the concept of UCAV operator training, which calls for operators to train at simulators instead of using the actual vehicles. A third concept that must be validated is the reduced operator-to-aircraft ratio. Developers believe that operators can control from four-to-six UCAVs at a time, which is made possible by the amount of autonomy that UCAVs will possess. All of these concepts will contribute to lowering O&S costs and increasing the cost-effectiveness of UCAVs, but there is a risk that some or all of these concepts will prove infeasible. The feasibility of these concepts depends to a great extent on the technology that becomes available and whether that technology can be integrated into a UCAV system.

Another challenge to UCAVs is presented by competing weapons systems. Cruise missiles, manned aircraft, and space assets all present possible challenges to UCAVs because all could conceivably accomplish the missions for which UCAVs are slated. UCAVs promise to be more flexible and cost-effective than cruise missiles, less risky (regarding friendly casualties) and cheaper than manned aircraft, and cheaper and less complex (technologically and politically) than space-based systems. There are, however, significant advantages that cruise missiles and manned aircraft have over UCAVs. Both are proven technologies and require less technological risk than UCAVs, and both have already been incorporated into the US force structure. There is concern that UCAVs will be difficult to integrate with manned aircraft and concern as to whether or not the required technology will be available and mature. Though it is promising, it remains to be seen how well UCAV technology will develop. Moreover, the challenge of integrating UCAVs into the force structure is an important issue that must be addressed before an accurate judgement on UCAVs can be made.

Lack of service cooperation is currently not a concern because the Air Force is the only service working extensively to develop UCAVs. The Navy has investigated the idea, but right now it is in a wait-and-see posture. If DARPA and the Air Force prove UCAVs to be feasible, it is likely that the Navy will investigate the possibilities further. Representative Hunter points out some of the downfalls of joint programs: "The genius of jointness, I have learned, . . . is not when everybody gets their input, but the genius of it is having somebody who, when everybody has made their input, makes the cut and says, this is what we do, and we move out. If you end up with jointness being everybody has a veto, or you end up with a weapon system that looks like it was built by a committee, we are generally in big trouble."[239] The USAF and DARPA should continue to lead the program; and if the time comes to include the Navy, Representative Hunter's advice should be heeded.

Effective management is a positive dimension of the current UCAV ATD program. The degree of user involvement in the program is significantly greater than in other DARPA UAV programs. Lessons from the past are what prompted this increased involvement, and the intent was to ensure that the end product met the needs of the user.[240] The challenge will be to continue this involvement throughout the program; and, if it continues to work out well, to use it as a benchmark for subsequent ATDs and ACDTs.

A UCAV demonstrator system, plans for a UCAV operational system, and risk reduction activities are the products that should come out of the UCAV ATD which will cost a total of $140 million in taxpayer money; and a low return on investment will generate political concern. As noted above, UAVs have generated many questions in Congress over the past 20 years, leaving many members wondering why so much money goes into UAV technology with so little return to the war fighter. Congress appears to have an interest in obtaining a solid UAV capability, but failed programs like the recently canceled DarkStar will only hurt congressional support. It is important that UAVs and UCAVs maintain a good track record with Congress if they are to receive the funding necessary to achieve operational capability.

Many people believe that USAF pilots are reluctant to accept UCAVs into the Air Force because of a pro-pilot bias or the white scarf syndrome. They believe that the Air Force pilot's true affection is not for airpower theory and combat effectiveness but for the airplane itself. Thus, USAF leaders, who are mostly pilots, would never accept an alternative method of achieving airpower. Though it is difficult to prove what really motivates people in their decision making, there is evidence that contradicts the white scarf syndrome. USAF leaders who broke the mold of this syndrome are Generals LeMay, Schriever, Fogleman, and Kostelnik. All of these men were pilots and all held high-level positions. They also expressed support for methods other than airplanes to achieve airpower. Though these are only a few examples, they suggest that perhaps there is another explanation why alternative forms of airpower have difficulty becoming a part of the USAF force structure. This study suggests that the uncertainty and risk surrounding systems such as missiles and UCAVs erodes the Air Force's support for these weapons when they are introduced. Where the Air Force has built its faith in airplanes, a new system like UCAV has to be proven before it will be accepted to carry out missions currently performed by manned aircraft. The risks and uncertainties have to be properly managed in order for UCAVs to receive the support necessary to realize their full potential.

Recommendations

The potential payoff of UCAVs is high and is a technology worth exploring. Though there are some risks involved and some challenges to be overcome, none of the challenges appear insurmountable. Since it is an ATD and not an ACTD, it does not have to become operational until it has proven itself technologically. The investment in the ATD, which will be $140 million, is what is at stake in exploring UCAVs; and the possible return on investment is priceless. To give UCAVs the best opportunity to succeed as a weapon system, this study offers the following recommendations to help manage the risks, reduce the uncertainties, and garner support for UCAVs:

- UCAVs should operate in a semiautonomous mode; there must be a man in the loop to authorize weapons delivery.

The amount of system autonomy is a theme that affects several aspects of UCAV development. The more autonomous the system, the more complex and expensive the technology to research and develop such a system. On the other hand, with more autonomy there are fewer C^2 requirements necessary to accommodate the interface between the operator and the UCAV. Fewer controllers are required for a more autonomous system because the system operates independently of controller input. These technological aspects impact cost-effectiveness which, in turn impacts how well UCAVs compete with other weapons systems. However, despite the advantages of operating UCAVs fully autonomously, UCAVs should be operated in a semiautonomous mode to help reduce risk and uncertainty.

A critical factor that relates directly to the issue of autonomy is public accountability. Recent history has shown that the American public and the international community hold military organizations and their members accountable for accidents during times of peace and war.[241] With worldwide television coverage allowing images of war to be broadcast into homes almost real time, the public has also grown sensitive even to the legitimate use of lethal force in war. "Technology has legitimized precision warfare, and 'criminalized' collateral death and destruction resulting from the use of lethal force. The perception exists among the press and public that it is now possible to prevent nearly all types of accidents and mistakes and only shoot the 'bad guy.'"[242]

These perceptions place limits on the use of any system that can deliver lethal force. UCAVs fall into this category, and there must be accountability designed into the system. UCAVs should be allowed to fly the entire mission autonomously and search for and detect the target autonomously, but a man in the loop must be there to monitor the system and give consent for weapons delivery. UCAVs will be scrutinized if and when an accident occurs, especially if the accident results in the inadvertent loss of life or treasure. The public will demand answers; therefore, the design, development, and employment of UCAVs must integrate the concept of accountability. Having a man in the loop will reduce the risk of mechanical and computer failures that result in the accidental employment of lethal weapons, and it will reduce uncertainty as to who is responsible and accountable when accidents do occur. This is particularly important for a system that is supposed to do a great deal of thinking on its own. "Humans must remain in the C^2 loop, and the internal and external systems and links must be robust enough to keep that loop intact. The sociopolitical implications are too high to ignore these facts."[243] Public and political support for UCAVs will erode if an accident occurs with no accountability. Operating UCAVs semiautonomously will also have implications regarding the types of missions they fly.

- Semiautonomous UCAVs should not be integrated into strike packages with manned aircraft.

In fulfilling the SEAD requirement for 2015, some visionaries see UCAVs integrating with manned aircraft to protect strike packages from the enemy SAM threat. In this type of SEAD, the engagement time line is unpredictable and can be very short. There is very little time to respond to the threat and even less time for error. In this case an autonomous UCAV would be necessary because the engagement time line may be too short for the UCAV to gain consent from a man in the loop. A semiautonomous UCAV that requires input from a remote operator runs the risk of C^2 failure or delay, and this could be deadly to a strike package. A UCAV used to protect a strike package would also have to employ mature and robust ATR technology to ensure the highest confidence in destroying the correct target without man-in-the-loop confirmation. This is not only important from the standpoint of limiting collateral damage but also because the UCAV's priority must be to protect the strike package. It cannot afford to expend time and weapons on anything except targets that pose a threat.

A more appropriate mission for a semiautonomous UCAV is an independent, destructive SEAD mission, sometimes referred to as DEAD (destruction of enemy air defenses). In this role the UCAVs fly search-and-destroy missions independent of manned strike packages, and they go out at a time and place of the planners' choosing. They search for threats, and time is less critical because they are neither integrated nor synchronized with a strike package and are not responsible for protecting one. Planners can also take advantage of the UCAV's ability to fly long endurance missions, which allows them to search the target area for extended periods of time. There is also time to obtain consent for weapons delivery or to reestablish communications links in case of C^2 failures without jeopardizing entire strike packages. The only danger that may arise in case of delayed consent is to the UCAV itself, which results in no loss of life.

There are other advantages to using semiautonomous UCAVs in an independent, destructive SEAD role. Since the UCAV will not be in the target area with the rest of the strike package, the burden of integrating and deconflicting the UCAV with several manned aircraft is eliminated. There may also be fewer C^2 problems because the UCAVs do not have to share limited bandwidth with the rest of a large strike package. Reducing risk and uncertainty by initially limiting the role of UCAVs will help build support for the new weapon in the Air Force community. The biggest disadvantage to limiting UCAVs in this way is that their full capability is not exploited. But as UCAVs mature operationally and technologically and as Air Force leaders and aviators gain confidence in their effectiveness, the role of UCAVs and the degree of autonomy they enjoy can be expanded. The final recommendation is directed toward achieving technological maturity that UCAVs require.

- Rely more on civilian institutions to develop and mature the technologies critical to the success of UCAVs.

During the 1960s the Air Force used the Big Safari program to research and develop the Ryan 147, one of the most successful UAV programs in history. Big Safari was a nontraditional approach to research and development that was shorter and less restrictive than normal methods. UCAV development is currently faced with many technological challenges, including the need for more robust bandwidth and data-link capability and more effective artificial intelligence technologies. There is also a challenge to mature existing technologies and integrate them into a UCAV operating system. Developing and maturing the necessary technologies will allow UCAVs to be proven operationally, to compete with other weapon systems, and to earn political and military support. Though DARPA and the Air Force are looking to the Boeing Company for some of these answers, there are many more resources in the civilian sector that could be tapped into for solutions.

Industry and universities are where much of our nation's brainpower resides, and the Air Force and DARPA should take advantage of this resource to help with the technological challenges regarding UCAVs. Although this is a nontraditional way of conducting military research and development, it is not new. As commander of the Army Air Corps before World War II, Gen Henry "Hap" Arnold was of the opinion that, "If the Air Corps had little money for research and development, then perhaps universities and industry could be persuaded to find some."[244] To provide incentive for research, "Arnold had very cleverly linked Air Corps development to civilian prosperity in the aviation industry, hoping civilian institutions would pick up the fumbled research ball while the Air Corps was struggling to acquire planes."[245] When General Arnold took command of the Air Corps in 1938, many research and development projects were under way including radar, aircraft windshield deicing, the jet-assisted takeoff system, and several aircraft and engine design modifications. "Many of these projects were related to the brand new B-17, an aviation leap in itself. Arnold wasted no time calling the 'long hairs' to a meeting at the National Academy of Sciences (NAS) under the auspices of the Committee on Air Corps Research, to solve these problems."[246] When asked why he was associating with these scientists and academics, "Arnold replied, that he was 'using' their brainpower to develop devices 'too difficult for the Air Force engineers to develop themselves.'"[247]

Many of the technological challenges facing UCAVs today have existed since the 1970s. The C^2 difficulties that plagued the BGM-34A during the Vietnam era are also potential problems for tomorrow's UCAVs. Perhaps these and other technological problems could be passed on to civilian institutions where some of our nation's brightest people could grapple with them. Instead of being limited to Air Force scientists or a few industry teams, the problem should be made available to anyone who wants to take a crack at it. Much of this technology, particularly the C^2 technologies like bandwidth, are applicable in the civilian sector and could be linked to civilian prosperity. Using innovative approaches to research and development may produce innovative answers to technological challenges. Breaking away from traditional methods, as "Hap" Arnold did, may help make the UCAV as operationally successful in 2015 as the B-17 was in World War II.

UCAV is a promising technology; and though there are some risks and uncertainties involved, the potential payoffs are high. There are technological as well as operational risks and uncertainties that must be properly managed for the UCAV concept to receive the support it needs to overcome the challenges that stand between it and operational capability. Other countries—including France, Great Britain, and Israel—are exploring the possibilities of UCAVs.[248] Air Vice Marshal R. A. "Tony" Mason defines airpower as "the exploitation of the third dimension *by* man, not necessarily *with* man."[249] If the USAF does not at least explore the possibilities, it may lose some of the airpower edge that it has enjoyed for so many years. After all, it was Giulio Douhet, one of the great airpower pioneers, who wrote, "Victory smiles upon those who anticipate the changes in the character of war, not upon those who wait to adapt themselves after the changes occur."[250]

Notes

1. Bill Sweetman, "Fighters without Pilots," *Popular Science*, November 1997, 96.
2. Tom Ehrhard, "The US Air Force and Unmanned Aerial Vehicles" (PhD diss., Johns Hopkins University, 1999), 3.
3. William Wagner and William P. Sloan, *Fireflies and Other UAVs* (Arlington, Tex.: Midland Publishing, 1992), ix.
4. Bruce W. Carmichael et al., "StrikeStar 2025" Air University Report, August 1996; on-line, Internet, 15 January 1999, available from http://www.au.af.mil/au/2025/volume3/chap13/v3c13-6.htm, 2-3.
5. Wagner and Sloan, 15.
6. Clive Hart, *Kites: An Historical Survey* (Mount Vernon, N.Y.: Paul P. Appel, 1982), 25.
7. Ehrhard, 3.
8. Ibid., 3.
9. Hart, 147.
10. Ehrhard, 7.
11. Ibid., 8.
12. Hugh McDaid and David Oliver, *Smart Weapons* (New York: Barnes and Noble, 1997), 10.

13. Kenneth P. Werrell, *The Evolution of the Cruise Missile* (Maxwell Air Force Base [AFB], Ala.: Air University Press, 1985), 8.
14. Ibid.
15. Ibid., 12.
16. Ibid.
17. Wagner and Sloan, 15.
18. Werrell, 14.
19. Wagner and Sloan, 15.
20. Werrell, 16–17.
21. Ibid., 16.
22. Wagner and Sloan, 15.
23. McDaid and Oliver, 16.
24. Ibid., 11.
25. Ibid.
26. Ibid.
27. Ibid.
28. Wagner and Sloan, 15.
29. Werrell, 32–35.
30. Ibid., 34.
31. McDaid and Oliver, 13.
32. Wagner and Sloan, 15.
33. Ibid.
34. Ibid., 16.
35. Ibid., 1.
36. William Wagner, *Lightning Bugs and Other Reconnaissance Drones* (Fallbrook, Calif.: Aero Publishers, 1982), ix.
37. Ibid.
38. Wagner and Sloan, 1.
39. Ibid., 2.
40. Ibid.
41. Ibid.
42. Ibid., 3
43. Ibid., 4.
44. Ibid., 6.
45. Ibid., 9.
46. Ibid., 10.
47. Ibid., 9.
48. Ibid.
49. Ibid.
50. Ibid.
51. Ibid., 11.
52. Ibid.
53. Ibid.
54. Ibid., 35.
55. Ibid., 36.
56. McDaid and Oliver, 42.
57. Ibid., 46.
58. Ibid., 42.
59. Wagner and Sloan, 38.
60. McDaid and Oliver, 46.
61. Wagner and Sloan, 40.

62. Ibid., 38.
63. Ibid., 42–43.
64. McDaid and Oliver, 46.
65. Ibid., 46.
66. Wagner and Sloan, 47.
67. James W. Plummer, "Temper RPV Enthusiasm," *Aviation Week & Space Technology*, 23 June 1975, 7.
68. Wagner and Sloan, 47.
69. Ibid., 96.
70. Wagner, 173.
71. Ibid., 174.
72. Ibid.
73. Ibid.
74. Wagner and Sloan, 96.
75. Wagner, 174.
76. Ibid., 176.
77. Wagner and Sloan, 97.
78. Ibid.
79. Wagner, 179.
80. Clarence A. Robinson, "Harpoon Slated for Prime Anti-Ship Role," *Aviation Week & Space Technology*, 7 May 1973, 16–19.
81. Ibid., 19.
82. Wagner, 179.
83. Wagner and Sloan, 98.
84. Wagner, 180.
85. Ibid., 181.
86. Ibid., 183.
87. Ibid., 182.
88. Wagner and Sloan, 96.
89. Wagner, 182.
90. Wagner and Sloan, 99.
91. Ibid.
92. Wagner, 183.
93. Ibid., 184–85.
94. Steven M. Shaker and Alan R. Wise, *War without Men: Robots on the Future Battlefield* (Washington, D.C.: Pergamon-Brassey's, 1988), 34.
95. Wagner, 185.
96. "Decision by Pentagon on RPVs for Combat Expected in Months," *Aviation Week & Space Technology*, 19 June 1972, 49.
97. History, Tactical Air Command, July 1972–June 1973, document 44; message from the chief of staff of the Air Force to Tactical Air Command, subject: Tactical Drone Capability. This document is located at the Air Force Historical Research Agency at Maxwell AFB, Ala.
98. Wagner and Sloan, 100–101.
99. Ibid.
100. Ibid., 101.
101. Ibid.
102. Ibid., 105.
103. Ibid.
104. Ibid., 106.
105. Ibid., 107.

106. Ibid.
107. Ibid., 108.
108. Ibid., 107–8; and McDaid and Oliver, 143, 148.
109. Wagner and Sloan, 108.
110. Ibid.
111. Randall Johnson, "Limitations on the Acquisition of RPVs: Technical, Political, or Managerial?" (Maxwell AFB, Ala.: Air Command and Staff College, 1975), 30.
112. Lionel Van Deerlin, "On Schedule, Within Budget," *Congressional Record*, vol. 120 (24 July 1974): 807.
113. Johnson, 23–24.
114. Van Deerlin, 807.
115. House, Committee on Appropriations, *Report: Department of Defense Appropriations Bill, 1974*, 93d Cong., 1st sess., 1973, Report No. 93-662, 216.
116. Barry Miller, "RPVs Provide US New Weapon Options," *Aviation Week & Space Technology*, 22 January 1973, 39.
117. Ibid., 40.
118. "National RPV Policy Needed?" *Armed Forces Journal*, February 1973, 20.
119. Johnson, 46.
120. Plummer, 7.
121. "Air Force's Plummer Notes Problems in Acceptance of RPVs," *Aerospace Daily*, 3 September 1975, 13.
122. House, Committee on National Security, Military Procurement Subcommittee, *Unmanned Aerial Vehicles Program*, 105th Cong., 1st sess., 9 April 1997, 902.
123. McDaid and Oliver, 60.
124. House, Committee on National Security, 902.
125. David A. Fulghum, "UAVs Pressed into Action to Fill Void," *Aviation Week & Space Technology*, 19 August 1991, 59.
126. Ibid.
127. McDaid and Oliver, 60.
128. House, Committee on National Security, 832.
129. Ibid., 902.
130. McDaid and Oliver, 60.
131. Ibid., 78.
132. John A. Tirpak, "The Robotic Air Force," *Air Force Magazine*, September 1997, 74.
133. Defense Advanced Research Projects Agency (DARPA), *Unmanned Combat Air Vehicle Advanced Technology Demonstration (UCAV ATD)*, 9 March 1998, on-line, Internet, 10 December 1998, available from http://www.darpa.mil/tto/ucav/ucav-sol.html.
134. Ibid.
135. Ibid.
136. Mike Leahy, DARPA, interviewed by author, 5 March 1999.
137. Robert Wall, "Boeing Wins UCAV Contract," *Aviation Week & Space Technology*, 29 March 1999, 84–85.
138. Wes Kremer, *Update on UCAV Activities*, briefing, ACC/DRA.
139. Brian P. Tice, "Unmanned Aerial Vehicles, Force Multiplier of the 1990s," *Airpower Journal*, Spring 1991, 53.

140. David A. Fulghum, "Payload, Not Airframe, Drives UCAV Research," *Aviation Week & Space Technology*, 2 June 1997, 52.

141. Ibid.

142. Mark Walsh, "Battle Lab Starts Study of Drones That Can Kill," *Air Force Times*, 28 July 1997, 27.

143. Jim Shane, UAV Battle Lab, Eglin AFB, Fla., interviewed by author, 22 February 1999.

144. John A. Tirpak, "UCAVs Move toward Feasibility," *Air Force Magazine*, March 1999, 34.

145. Ibid.

146. Leahy interview.

147. Ibid.; and Society of Wild Weasels home page, December 1997, n.p.; on-line, Internet, 24 April 1999, available from http://www.home1.gte.net/weasels/index.htm. "The Wild Weasel mission was developed by the U.S. Air Force about 30 years ago, during the Vietnam War era. Its primary concept was the use of two-seat aircraft, to counter hostile radar-controlled surface-to-air weapons. The first Wild Weasel aircraft were F-100Fs. Next came F-105s, followed by F-4Cs. The 'last of the breed' were the F-4Gs. . . . The Weasel role has now been passed on to a 'new breed' of animal: a single-seat F-16 'Viper.'"

148. Leahy interview.

149. DARPA, *Systems Capability Document (SCD)*, 9 March 1998; and on-line, Internet, 10 December 1998, available from http://www.darpa.mil/tto/ucav/ucavappen.html, 1.

150. Greg Jenkins, Air Armament Center–Armament Product Group Manager's Office, Eglin AFB, Fla., interviewed by author, 23 February 1999.

151. Ibid.

152. Ibid.

153. Ibid.

154. "Air Force 2025: Executive Summary," 2025 Support Office, Air University, August 1996; and on-line, Internet, 10 December 1998, available from http://www.au.af.mil/au/2025/monographs/E-s/e-s.htm, 33.

155. Carmichael et al., "StrikeStar 2025," 6-6.

156. Ibid., 8.

157. Ibid.

158. Michael Kostelnik, commander, Air Armament Center, Eglin AFB, Fla., interviewed by author, 24 February 1999.

159. DARPA, *Unmanned Combat Air Vehicle Advanced Technology Demonstration (UCAV ATD)*.

160. Tirpak, "UCAVs Move toward Feasibility," 37.

161. Leahy interview.

162. Fred Zayas, ACC/DR Advanced Programs Office, interviewed by author, 11 March 1999.

163. "UAVs Take Off into a Multifunction Future," *Jane's Defence Weekly*, 12 August 1995, 33.

164. House, Committee on National Security, 831.

165. James Elliot, "UCAVs: Towards a Revolution in Air Warfare?" *Military Technology*, August 1998, 17.

166. Sweetman, 98.

167. Ibid.

168. David R. Mets, "Air Armament Technology for the Deep Attack: Did It Work? What If It Works Next Time?" (Maxwell AFB, Ala.: School of Advanced Airpower Studies, 4 March 1999), n.p.

169. Rich Mook, small smart bomb team leader, Air Force Research Lab Munitions Directorate, Eglin AFB, Fla., interviewed by author, 23 February 1999.

170. Ibid.

171. Kenneth Edwards, LOCAAS team leader, Air Force Research Lab Munitions Directorate, Eglin AFB, Fla., interviewed by author, 23 February 1999.

172. Ibid.

173. Ibid.

174. Kostelnik interview.

175. Leahy interview.

176. Elliot, 18.

177. Ibid.

178. Shane interview.

179. Zayas interview.

180. Leahy interview.

181. Zayas interview.

182. Tirpak, "UCAVs Move toward Feasibility," 34.

183. Tirpak, "The Robotic Air Force," 71.

184. Sweetman, 99.

185. Leahy interview.

186. Tirpak, "UCAVs Move toward Feasibility," 34.

187. Ibid.

188. Leahy interview.

189. Tirpak, "UCAVs Move toward Feasibility," 35.

190. Leahy interview.

191. Paul Geier, chief of the Combat Applications Division of the UAV Battle Lab, interviewed by author, 22 February 1999.

192. Ibid.

193. Peter J. Worch, *UAV Technologies and Combat Operations*, vol. 1, *Summary*, United States Air Force Scientific Advisory Board, November 1996, 9-5.

194. Leahy interview; and David S. Fadok, "John Boyd and John Warden: Airpower's Quest for Strategic Paralysis," in *The Paths of Heaven: The Evolution of Airpower Theory*, ed. Phillip S. Meilinger (Maxwell AFB, Ala.: Air University Press, 1997), 366. The observation, orientation, decision, and action (OODA) loop is a model developed by John Boyd that depicts a decision-making cycle. In Boyd's OODA loop concept, "the crux of winning becomes the relational movement of opponents through their respective OODA loops. Whoever repeatedly observes, orients, decides, and acts more rapidly (and accurately) than his enemy will win."

195. Carmichael et al., 4-3.

196. Ibid.

197. Ibid.

198. House, Committee on National Security, 915.

199. David A. Fulghum, "Navy Wants UCAVs for Carrier Use," *Aviation Week & Space Technology*, 2 June 1997, 55.

200. Graham Warwick, "UCAVs Head to Sea," *Flight International*, 14–20 October 1998, 61.

201. Leahy interview.

202. Tirpak, "UCAVs Move toward Feasibility," 37.

203. Warwick, 61.

204. House, Committee on National Security, 833. The cancellation of DarkStar and the fielding of Predator are included in these numbers.

205. Tirpak, "UCAVs Move toward Feasibility," 34.

206. Leahy interview.

207. Zayas interview.

208. McDaid and Oliver, 107.

209. Zayas interview.

210. Leahy interview.

211. Ibid.

212. Zayas interview.

213. Tirpak, "UCAVs Move toward Feasibility," 37.

214. Zayas interview.

215. Worch, 9-3.

216. Ibid., 9-4.

217. Ibid.; and "Northrop AGM-136A 'Tacit Rainbow' Unmanned Aerial Vehicle," n.p.; on-line, Internet, 17 April 1999, available from http://www.ipcress.com/writer/area51/rainbow.html. The precedent comes from the exclusion of the Norththrop AGM-136A Tacit Rainbow. The Tacit Rainbow was conceived in the early 1980s, and its purpose was to supplement manned aircraft in the SEAD mission. The air-launched AGM-136A flew to its target area and loitered until it sensed transmissions from an enemy radar. It then attacked by flying into the radar site, destroying the target with its 40-pound warhead. Many considered the Tacit Rainbow a UCAV, but by this study's definition it was not because it did not return after completing its mission. The program was cancelled in 1991 due to budgetary constraints, but it was specifically excluded from the START treaty before the program ended.

218. House, Committee on National Security, 904.

219. Ibid., 861.

220. Ibid., 827.

221. Ibid., 851.

222. Tice, 52.

223. Carl Builder, *The Icarus Syndrome: The Role of Air Power Theory in the Evolution and Fate of the US Air Force* (New Brunswick, N. J.: Transaction Publishers, 1994), 35.

224. Peter Grier, "DarkStar and Its Friends," *Air Force Magazine*, July 1996, 41.

225. Edmund Beard, *Developing the ICBM* (New York: Columbia University Press, 1976), 39.

226. "Reaching for the Stars," *Air Force News*, two pages; and on-line, Internet, 15 March 1999, available from http://www.af.mil/news/airman/0298/pioneer.htm, 1.

227. "Father of the Air Force Space Program," one page; and on-line, Internet, 15 March 1999, available from http://www.zainet.com/space/schriever.html, 1.

228. "Reaching for the Stars," 1.

229. Ibid.

230. Kostelnik interview.

231. Walton S. Moody, *Building a Strategic Air Force* (Washington, D.C.: Air Force History and Museums Program, 1996), 423–24.

232. Plummer, 7.

233. Zayas interview.

234. Stephen P. Rosen, *Winning the Next War* (Ithaca, N.Y.: Cornell University Press, 1991), 249–50.

235. Monica Munro, "UAV Pilot Receives Air Medal," ACC News Service, 26 November 1997, n.p.; and on-line, Internet, 2 April 1999, available from http://www.acc.af.mil/news/nov/970229.html.

236. Kostelnik interview.

237. Clive Maxfield and Alvin Brown, "1973 AD to 1981 AD: The First Personal Computers," *Bebop Bytes Back: An Unconventional Guide to Computers* (Madison, Ala.: Doone Publications, 1998), n.p.; and on-line, Internet, 2 April 1999, available from the frame page for *A History of Computers*, http://www.maxmon.com/history.htm.

238. David A. Fulghum, "Will New Elusive Craft Rise From DarkStar?" *Aviation Week & Space Technology*, 22 February 1999, 28.

239. House, Committee on National Security, 862.

240. Leahy and Zayas interviews.

241. Carmichael et al., 4-5. There are two examples of how the public will respond to accidents or mistakes in applying lethal force. First is the shootdown of two US helicopters by Air Force fighters supporting Operation Provide Comfort in northern Iraq resulting in the loss of 24 lives. A second example is the 1998 incident in which a Marine A-6 clipped a gondola wire in Italy that caused the deaths of 20 people. In both cases individuals were held accountable for their actions.

242. Ibid.

243. Ibid.

244. Dik Daso, "Origins of Airpower: Hap Arnold's Command Years and Aviation Technology, 1936–1945," *Airpower Journal*, Fall 1997.

245. Ibid.

246. Ibid., 96.

247. Ibid., 98.

248. Sharon Sadeh, "Israel's UAV Industry Seeking New Frontiers," *Military Technology*, June 1995, 15; Steven Zaloga, "UAVs Gaining Credibility," *Aviation Week & Space Technology*, 12 January 1998, 95; and Robert Williams, "Air Force Vision Anticipates Victory Via New Technology," *National Defense*, March 1997, 17.

249. Air Vice Marshal R. A. "Tony" Mason, *Air Power: A Centennial Appraisal* (Washington, D.C.: Brassey's, 1997), 276.

250. *Contrails: The Air Force Cadet Handbook*, United States Air Force Academy, vol. 28, 1982, 154.

Bibliography

Books

Beard, Edmund. *Developing the ICBM.* New York: Columbia University Press, 1976.

Builder, Carl H. *The Icarus Syndrome: The Role of Air Power Theory in the Evolution and Fate of the US Air Force.* New Brunswick, N.J.: Transaction Publishers, 1994.

Fadok, David S. "John Boyd and John Warden: Airpower's Quest for Strategic Paralysis." In *The Paths of Heaven: The Evolution of Airpower Theory.* Edited by Phillip S. Meilinger. Maxwell AFB, Ala.: Air University Press, 1997.

Hart, Clive. *Kites: An Historical Survey.* Mount Vernon, N.Y.: Paul P. Appel, 1982.

Mason, Air Vice Marshal R. A. "Tony." *Air Power: A Centennial Appraisal.* Washington, D.C.: Brassey's, 1997.

McDaid, Hugh, and David Oliver. *Smart Weapons.* New York: Barnes and Noble, 1997.

Moody, Walton S. *Building a Strategic Air Force.* Washington, D.C.: Air Force History and Museums Program, 1996.

Rosen, Stephen P. *Winning the Next War.* Ithaca, N.Y.: Cornell University Press, 1991.

Shaker, Steven M., and Alan R. Wise. *War without Men: Robots on the Future Battlefield.* Washington, D.C.: Pergamon-Brassey's, 1988.

Taylor, John W. R. *Jane's Pocket Book of Remotely Piloted Vehicles.* New York: Macmillan, 1977.

United States Air Force Academy. *Contrails: The Air Force Cadet Handbook*, 1982.

Wagner, William. *Lightning Bugs and Other Reconnaissance Drones.* Fallbrook, Calif.: Aero Publishers, 1982.

Wagner, William, and William P. Sloan. *Fireflies and Other UAVs.* Arlington, Tex.: Midland Publishing, 1992.

Werrell, Kenneth P. *The Evolution of the Cruise Missile.* Maxwell Air Force Base (AFB), Ala.: Air University Press, 1985.

Internet

"Air Force 2025: Executive Summary." 2025 Support Office, Air University, August 1996. On-line. Internet, 10 December 1998. Available from http://www.au.af.mil/au/2025/monographs/E-s/e-s.htm.

Carmichael, Bruce W., et al. "StrikeStar 2025" Air University Report, August 1996, n.p. On-line. Internet, 15 January 1999. Available from http://www.au.af.mil/au/2025/volume3/chap13/v3c13-6.htm.

Defense Advanced Research Projects Agency (DARPA). *Systems Capability Document (SCD)*, 9 March 1998, On-line. Internet, 10 December 1998. Available from http://www.darpa.mil/tto/ucav/ucavappen.html.

———. *Unmanned Combat Air Vehicle Advanced Technology Demonstration (UCAV ATD)*, 9 March 1998, On-line. Internet, 10 December 1998. Available from http://www.darpa.mil/tto/ucav/ucav-sol.html.

"Father of the Air Force Space Program," one page. On-line. Internet, 15 March 1999. Available from http://www.zainet.com/space/schriever.html.

Maxfield, Clive, and Alvin Brown. "1973 AD to 1981 AD: The First Personal Computers." *Bebop Bytes Back: An Unconventional Guide to Computers*. Madison, Ala.: Doone Publications, 1998, n.p. On-line. Internet, 2 April 1999. Available from the frame page for the *History of Computers*, http://www.maxmon.com/history.htm.

Munro, Monica. "UAV Pilot Receives Air Medal." ACC News Service, 26 November 1997, n.p. On-line. Internet, 2 April 1999. Available from http://www.acc.af.mil/news/nov/970229.html.

"Northrop AGM-136A 'Tacit Rainbow' Unmanned Aerial Vehicle." n.p. On-line. Internet, 17 April 1999. Available from http://www.ipcress.com/writer/area51/rainbow.html.

"Reaching for the Stars." *Air Force News*, two pages. On-line. Internet, 15 March 1999. Available from http://www.af.mil/news/airman/0298/pioneer.htm.

Society of Wild Weasels home page, December 1997, n.p. On-line, Internet, 24 April 1999. Available from http://home1.gte.net/weasels/index.htm.

Papers

Ehrhard, Tom. "The Genesis of UAVs." PhD diss., Johns Hopkins University, 1999.

Fadok, David S. "John Boyd and John Warden: Airpower's Quest for Strategic Paralysis." In *The Paths of Heaven: The Evolution of Airpower Theory*. Edited by Phillip Meilinger, Maxwell AFB, Ala.: Air University Press, 1997.

Fox, Roy. "UAVs: Holy Grail for Intel, Panacea for RSTA, of Much Ado about Nothing? UAVs for the Operational Commander." Unpublished paper, Naval War College, 13 February 1998.

Howard, Stephen P. *Special Operations Forces and Unmanned Aerial Vehicles: Sooner or Later?* Maxwell AFB, Ala.: Air University Press, February 1996.

Johnson, Randall. "Limitations on the Acquisition of RPVs: Technical, Political, or Managerial?" Maxwell AFB, Ala.: Air Command and Staff College, 1975.

Mets, David R. "Air Armament Technology for the Deep Attack: Did It Work? What if It Works Next Time?" Maxwell AFB, Ala.: School of Advanced Airpower Studies, 4 March 1999.

Renehan, Jeffrey N. *Unmanned Aerial Vehicles and Weapons of Mass Destruction: A Lethal Combination?* Maxwell AFB, Ala.: Air University Press, August 1997.

Periodicals

"Air Force's Plummer Notes Problems in Acceptance of RPVs." *Aerospace Daily*, 3 September 1975, 13–14.

Conley, Kathleen M. "Campaigning for Change." *Airpower Journal*, Fall 1998, 54–69.

Daso, Dik. "Origins of Airpower: Hap Arnold's Command Years and Aviation Technology, 1936–1945." *Airpower Journal*, Fall 1997, 95–111.

"Decision by Pentagon on RPVs for Combat Expected in Months." *Aviation Week & Space Technology*, 19 June 1972, 48–49.

Elliot, James. "UCAVs: Towards a Revolution in Air Warfare?" *Military Technology*, August 1998, 15–18.

Fulghum, David A. "Aircraft, UCAVs: An Uneasy Mix." *Aviation Week & Space Technology*, 3 August 1998, 68–70.

———. "Decades Are Needed to Perfect Unmanned War Planes." *Aviation Week & Space Technology*, 3 August 1998, 70–71.

———. "Gull-Wing UCAV Eyed for U.S. Aircraft Carriers." *Aviation Week & Space Technology*, 16 June 1997, 36–37, 58.

———. "Next Generation UCAVs Will Feature New Weapons and Engines." *Aviation Week & Space Technology*, 3 August 1998, 68–70.

———. "Navy Wants UCAVs for Carrier Use." *Aviation Week & Space Technology*, 2 June 1997, 55.

———. "Payload, Not Airframe, Drives UCAV Research." *Aviation Week & Space Technology*, 2 June 1997, 51–53.

———. "Pilots to Leave Cockpit in Future Air Force." *Aviation Week & Space Technology*, 5 February 1996, 26–28.

———. "UAVs Pressed Into Action To Fill Void." *Aviation Week & Space Technology*, 19 August 1991, 59–60.

———. "Unmanned Strike Next for Military." *Aviation Week & Space Technology*, 2 June 1997, 47–48.

———. "Will New Elusive Craft Rise From DarkStar?" *Aviation Week & Space Technology*, 22 February 1999, 27–28.

Gourley, Scott R. "Molding the Shape of Future Air Combat (Uninhabited Combat Air Vehicles)." *Jane's Defence Weekly*, 16 July 1997, 27.

Grier, Peter. "DarkStar and Its Friends," *Air Force Magazine*, July 1996, 40–45.

Miller, Barry. "RPVs Provide US New Weapon Options." *Aviation Week & Space Technology*, 22 January 1973, 39.

"Moulding the Shape of Future Combat." *Jane's Defence Weekly*, 16 July 1997, 27.

"National RPV Policy Needed?" *Armed Forces Journal*, February 1973, 19–20.

Owens, Mackubin Thomas. "Technology, the RMA, and Future War," *Strategic Review*, Spring 1998, 1–12.

Plummer, James W. "Temper RPV Enthusiasm," *Aviation Week & Space Technology*, 23 June 1975, 7.

Probert, Andrew A. "Uninhabited Combat Aerial Vehicles: Remove the Pilot?" *Airpower Journal*, Winter 1997, 85–89.

Roberts, Fred. "UCAV Update." *Air Combat*, October/November 1998, 39–41.

Roberts, William H. "Unmanned Combat Aircraft Age Is Rapidly Approaching: Aerospace Industry Launching Own Research, Development Initiatives." *National Defense*, January 1998, 22–23.

Robinson, Clarence A. "Harpoon Slated for Prime Anti-Ship Role." *Aviation Week & Space Technology*, 7 May 1973, 16–19.

Sadeh, Sharon. "Israel's UAV Industry Seeking New Frontiers," *Military Technology*, June 1995, 15–18.

Sweetman, Bill. "Fighters without Pilots." *Popular Science*, November 1997, 96–101.

———. "Green Light for UCAVs." *Interavia*, August 1998, 39.

Tice, Brian P. "Unmanned Aerial Vehicles: The Force Multiplier of the 1990s." *Airpower Journal*, Spring 1991, 41–54.

Tirpak, John A. "Complications Overhead (Intelligence, Surveillance, and Reconnaissance)." *Air Force Magazine*, April 1998, 22–28.

———. "The Robotic Air Force." *Air Force Magazine*, September 1997, 70–74.

———. "UCAVs Move Toward Feasibility." *Air Force Magazine*, March 1999, 32–37.

"UAVs Take Off into a Multifunction Future." *Jane's Defence Weekly*, 12 August 1995, 33–34, 37.

Wall, Robert. "Boeing Wins UCAV Contract." *Aviation Week & Space Technology*, 29 March 1999, 84–85.

Walsh, Mark. "Battle Lab Starts Study of Drones that Can Kill." *Air Force Times*, 28 July 1997, 27.

Warwick, Graham. "UCAVs Head to Sea." *Flight International*, 14–20 October 1998, 61.

Williams, Robert. "Air Force Vision Anticipates Victory Via New Technology." *National Defense*, March 1997, 16–17.

Zaloga, Steven. "UAVs Gaining Credibility." *Aviation Week & Space Technology*, 12 January 1998, 93–95.

Other

History. Tactical Air Command. July 1972–June 1973, Doc. 44.

House. Committee on Appropriations. *Report: Department of Defense Appropriations Bill, 1974*. 93d Cong., 1st sess., 1973; Report No. 93-662, 216.

———. Committee on National Security, Military Procurement Subcommittee. *Unmanned Aerial Vehicles Program*. 105th Cong. 1st sess., 9 April 1997.

Kremer, Wes. *Update on UCAV Activities*. Briefing. ACC/DRA.

Remotely Piloted Vehicles. Report of the Proceedings of the AFSC/RAND Symposium. vol. 2, August 1971.

Van Deerlin, Lionel. "On Schedule, Within Budget." *Congressional Record*, vol. 120, 24 July 1974, 806–7.

Worch, Peter J. *UAV Technologies and Combat Operations*, vol. 1, *Summary*, United States Air Force Scientific Advisory Board, November 1996.